STUBBORN & LIKING IT

STUBBORN & LIKING IT

EINSTEIN & OTHER GERMANS IN AMERICA

George F. Wieland

ANN ARBOR, MICHIGAN

Copyright © 2016 George F. Wieland

All rights reserved. Except as permitted under the U. S. Copyright Act of 1976, no part of this publication may be reproduced, distributed, transmitted in any form or by any means, or stored in a database or retrieval system, without the prior written permission of the author, George F. Wieland, Ann Arbor, MI 48104 or gwieland@umich.edu.

Chapter 7 is based on *Bessarabian Knight: A Peasant Caught between the Red Star and the Swastika* by I. Weiss and G. F. Wieland published by the American Historical Society of Germans from Russia, Lincoln, Nebraska, 1991.

On the cover: Einstein in Princeton, 1935, photo by Sophie Delar.

ISBN 13: 978-1518648472
ISBN-10: 1518648479

"If you want to live a happy life, tie it to a goal, not to people or objects."
Albert Einstein

CONTENTS

PREFACE xv

INTRODUCTION 1
   Stubbornness as Negative 2
   Stubbornness as Positive 2
   Intrinsic Motivation 4
   Problems with Extrinsic Motivations 5
   Needed Ingredients for Intrinsic Commitments 7
   Developing Intrinsic Motivation 9
   Escalation in Intrinsic Commitments 10
   From Extrinsic Motivation to Meaningfulness 11

1: CONRAD WEISER, PEACEMAKER 13
   Why They Left 13
   A Raw Deal 14
   Living with the Mohawks 15
   Another Raw Deal 16
   Interpreter for Pennsylvania 17
   Spiritual Life and Ephrata Cloister 18
   The Moravians 19
   Go-Between to the Iroquois 20
   New York Colony 22

    Trouble with the French 23

    Scalpings on the Frontier 23

    Hatred of the Indians 24

    Weiser Is Stripped of Power 25

    A Stubborn Triumph 26

2: COMMUNES **29**

    Rapp, a Charismatic Pietist 29

    Separatists 30

    Sacrifices 33

    Return to Pennsylvania 35

    The Ambassador of God 37

    Membership is Closed 38

    Enormous Wealth 40

    Swindler Duss 41

    Bethel and Zoar Communes 42

    Modifying the Harmony Model 44

3: FARMERS **47**

    Indians, Cattlemen, and Farmers 47

    Chain Migration 48

    Shock of Prairie Life 50

    Tough Women 52

    Making Alcohol 53

    A Visit to Twentieth Century Eustis 55

    Stubborn Germans 57

    Churchgoers 58

    Alcohol and More Alcohol 58

    Humor 60

Male Chauvinism 61

Changes 62

The Downside to Modernity 63

Too Much Drinking 65

4: EINSTEIN 69

Swabian Origins 69

Hell-Raiser 70

Childhood 72

Religion 73

Not Good at School 74

Miracle Year 75

Paradigm Shift: Relativity 76

Speed of Light as a Constant 77

Swabian Modesty Out of Place 79

Problems in High Society 81

Marriages 82

Women 84

Solving the Mystery of Gravity 86

Criticizing America 87

Pacifism 88

Against Nationalism 90

Refuge in America 91

A Gentle Clown? 92

Above All, Stubborn 93

The Dream of Subsuming Quantum Theory 94

Single-Minded to the End 95

Work 96

5: JEWS IN NAZI GERMANY 99

Forced Christianizing 99
Bavarian Expulsions 101
Life in the Villages 104
Jewish Germans 105
Government Officials Fired 107
Friends Change 109
Artistic Freedom Is Gone 110
The Nazi Vice Tightens 112
Social Life Ends 115
Night of Violence 115
Collaboration with the Nazis 118
The Gestapo 119
A Home Visit 121
Finally Ready to Leave 122
The Holocaust 123
Why Didn't They Leave Germany? 125
Leaving in Time 127
Safety in England 128
On Trial for Smuggling 130
Sad Leave-Taking 131
America: Pro and Con 133
Returning to Germany 134
To Israel? 137
Keeping the Old Customs 138
Acceptance of a German Heritage 139

6: PLAYWRIGHT BRECHT **143**

A Softie Inside *144*

Religious Origins of Brecht's Plays *145*

The Critic *146*

To Munich and Revolution *148*

Soviet Bavaria *150*

Hitler Appears *152*

The Appeal of Communism *154*

Committed to Communism *156*

The Nazis Take Charge *157*

Into Exile *160*

Hollywood *163*

Brecht's Enemies *165*

Failure on Broadway *166*

Political Problems *167*

A Bad Fit with American Theater *169*

Among Germans Again *171*

Formality and Privacy *172*

Brecht and His Women *174*

Tinkering in Berlin *176*

The Dialectic *177*

Communist East Germany *178*

7: BETWEEN HITLER AND STALIN **181**

Bessarabia *181*

Peasant Life *182*

Strictness *183*

Germany Was Different *184*

Weiss Becomes a Romanian Soldier	*185*
The Soviets Invade Bessarabia	*187*
Becoming a Deserter	*188*
Meeting the Russians	*189*
Under Communism	*190*
Rescued by Hitler	*190*
Resettled and Drafted	*192*
The Partisans	*195*
Fighting at the Front	*196*
Retreating	*197*
Wounded	*198*
The War Comes to Johanna Weiss	*199*
Under the Bombs	*201*
Johanna and the Russian Occupation	*202*
Immanuel Fights on the Western Front	*204*
Surrender	*204*
Chaos in Germany	*206*
Rescuing the Family	*207*
Time Zero	*208*
Starting Anew in Swabia	*209*
Leaving Germany	*211*
Trying to Farm Again	*213*
How We Did It	*214*
8: POSTWAR IMMIGRANTS	**217**
Sharing	*217*
Social Revolution	*219*
Political Awakening	*221*

    Schmidt and Working Class Life 225

    Becoming a Patriot.. 225

    The Bombing 227

    Fleeing Augsburg 230

    Enemy Soldiers 232

    The Military Occupation... 234

    A War Bride 236

    Impressions of America 237

    Bromke Speaks Out 239

    Stubborn Questioning of Religion 239

    Welfare from Hitler 241

    Geiger's Turning Point.. 242

    An Entrepreneur 243

    Making Americans Efficient 245

    The Super Shopper Plan 245

    On a Mission 247

9: CONCLUSIONS 249

    Development of Intrinsic Commitments... 249

    Decline of Intrinsic Commitments 253

    From Extrinsic to Intrinsic Commitment.. 254

    Intrinsic Motivation in Swabia 256

    Intrinsic Commitments in America? 260

SOURCES... 265

INDEX.. 287

THE AUTHOR 311

PREFACE

I am so grateful to my cousin, Julia Weber, for pointing in the window of a Murrhardt bookshop at Thaddäus Troll's *Deutschland deine Schwaben*. This book introduced me to the Swabians, the epitome of German stubbornness. My other relatives provided many memorable and interesting experiences, from serving me champagne at mid-morning to inviting me to *Stunde*, where devout Pietists met in the home after church.

The late Prof. Heinz Biesdorf of Cornell University and his wife Ellen entertained me over the years with stories about the distinctive culture of Swabia. I was also encouraged by visits to Prof. Ulrich Planck at the Universität Hohenheim, as well as Prof. Martin Scharfe, Prof. Hermann Bausinger, and the late Prof. Utz Jeggle, all at the Universität Tübingen, and the late Dr. Paul Sauer, historian and head of the state archives in Stuttgart.

Michael Betzold edited my manuscript, trying to fix my Germanic syntax. My readers over the years have been helpful not only in correcting drafts but also in providing encouragement: Dr. Barbara K. Petersen, Patience S. Wieland, Charlie Zakrajshek, David Lindemer, Dr. Elaine Hockman, and Sharon Kane Wieland. They are not responsible for the inevitable errors and infelicities that remain.

The University of Michigan and Ann Arbor District libraries obtained many books for me with interlibrary loans.

Finally, Nancy Shattuck of Wayne State University provided needed encouragement when I was overwhelmed with publication problems.

INTRODUCTION

Albert Einstein spent three decades fruitlessly working to combine electromagnetism and gravity into a unified field theory. His unfinished efforts were at his hospital bedside when he died at age 76. According to Einstein, the scientist's work is to find an important problem and not give up until success. He also claimed that he wasn't so smart. He just stayed with problems longer than others.

Bertolt Brecht was also stubborn. He worked endlessly trying to alert people to the harm that society did to the downtrodden and the poor. Even during a fourteen years in exile from the Nazis when he lacked access to German-speaking actors and audiences, he continued to write plays "for the drawer."

Other Germans have been similarly stubborn. Jews were so committed to being German citizens that they resisted Nazi abuse and Hitler's urging they leave. For five years, emigration to America was freely available, but they stubbornly continued to believe in their right to enjoy being Germans.

Conrad Weiser learned the Mohawk language as a teen in colonial America and spent the rest of his life mediating between Indians and colonists, even after losing his job and pay. Members of German nineteenth-century American religious communes stubbornly continued to await Christ's Second Coming, even after it didn't happen. German homesteaders in Nebraska continued farming despite many calamities like droughts, wild fires, and pestilence, while nearby Americans gave up and moved on.

Many people stereotype Germans as stubborn.[1,2] Germans

themselves have their own stereotype of a stubborn person—the Swabians of southwest Germany.[3-5] As usual, there is some truth to the stereotypes.[6] The stories here of Swabian Germans will confirm these stereotypes and explain the development of such stubbornness, including how it can lead to happiness.

(Superscript numbers point to sources listed at the end of the book. Appended page numbers in the superscripts give the location of quotations or numerical information.)

Stubbornness as Negative

American dictionaries define stubborn as refusing to yield, obey, or comply; resisting doggedly; and being unwilling to give in to a reasonable request. The word's origin comes from the Anglo-Saxon term for a tree stump, something that is resistant to removal.

The dictionary's negative view of stubbornness fits American society. Americans are high on the psychological trait of agreeableness.[7] They want to get along, to compromise, and to be flexible. As sociologist David Riesman famously put it, Americans are often "other directed."[8]

Some stubbornness is clearly like the negative definition in the dictionary. People can reject all requests for fear of losing options for freedom and independence.[9]

Such stubbornness may have made some Jews reluctant to leave Germany despite the increasing Nazi restrictions during the 1930s (Chapter 5). Many Jews were entrepreneurs who valued their independence. They didn't want the failed artist Hitler to tell them what to do or to leave. Some of them felt "hell no, we won't go."

Stubbornness as Positive

While some Germans may similarly reject all influence attempts, others are stubborn in another sense. In the German dictionary, stubbornness can mean E*igensinnig*, literally having "self-will." Stubbornness can come from a mind being fixed in a positive way. Most of the stories of Germans here will reflect a positive commitment to living and acting in a persistently single-minded way. These people are often happy, not distressed.

German immigrants in the Midwest had such a stubborn commitment to farming while the Irish, Polish, and others turned to alternative pursuits after they found the rewards in farming meager. The U.S. Census still shows German Americans overrepresented in rural census tracts. As a result, German Americans have lower average incomes than other ethnic groups choosing urban areas to live and work. If the Germans were seeking financial rewards, their stubbornness was not very useful. On the other hand, the German farmers' stubborn commitment enabled them to preserve important values—freedom and independence. In Chapter 3, we will find that farmers in Nebraska were stubborn because they liked farming as a way of life, for itself and not for rewards like money.

Immanuel Weiss (Chapter 7) also showed persistent single-mindedness. Two farms were taken from him, but despite many hardships, he never gave up his dream of again becoming an independent farmer. He left Germany for America and spent fifteen years in Iowa working as a farm laborer, sharecropper, and a farm renter before he could once more be fully independent and farm his own land. Weiss confirmed the saying, "you can take the boy off the farm, but you can't take the farm out of the boy." For Weiss and the Nebraska Germans, farming was part of their essential being, and it made them happy.

We can understand this single-mindedness as a form of commitment. The term "commitment" comes from the Latin "to bring together." An individual becomes one with a value. In religion, committed individuals have faith in God. Their beliefs aren't rationally weighed pro or con. People remain spiritually committed even when bad things happen. Marriage is another such commitment. One does not leave a marriage the minute problems arise.

Intrinsic Motivation

Important for understanding this kind of stubborn commitment is "intrinsic motivation." One feels that just doing it is satisfying. The activity and the process are rewarding in themselves. Often people prefer continuing such actions compared to behaviors aimed at achieving external rewards or avoiding punishments. Intrinsic satisfactions feel part of the individual and are thus meaningful. Einstein (Chapter 4) characterized this state of mind as "akin to that of the religious worshiper or the lover; the daily effort comes from no deliberate intention or program, but straight from the heart."[10,p363]

Such intrinsic motivation often involves total absorption. People lose themselves in the activity. Time flies enjoyably. [11]

There are several kinds of intrinsic motivation. First, there is intrinsic motivation to know, to get pleasure from learning, exploring, and understanding new things. Another form of intrinsic motivation is accomplishment, the satisfaction coming from mastery and trying to outdo oneself in creating or completing something. Finally, there is intrinsic motivation to experience stimulation and associated sensations. These activities can be ends in themselves, leading to stubborn persistence.[12]

In contrast, "extrinsic motivation" aims at getting something external to an individual, like money, praise, or avoiding punishment. Other external aims include dealing with threats, deadlines, directives, and imposed goals. The rewards of extrinsic motivation are separate from the means used. If such contingent rewards are removed for some reason, action often stops.[13]

Einstein contrasted extrinsic and intrinsic motivation when he claimed: "The sculptor, the artist, the musician, the scientist work because they love their work. Fame and honor are secondary." He added that "the mainspring of scientific thought is not an external goal toward which one must strive, but the pleasure of thinking."[10,pp10,364]

Of course, extrinsic rewards like fame or self-esteem can heighten continued persistence. However, when intrinsic motivation is the main push to action, stubborn doggedness can continue despite the removal of extrinsic rewards.[14] For example, someone committed to God, someone with faith, will continue praying even when prayers remain unanswered. They may feel, "It was God's will." On the other hand, people with extrinsic motivation to attend church—to socialize, meet important people, or show off fine clothing—will stop their attendance when these rewards are no longer present.

Intrinsic motivation often seems wrongheaded. A study of the World War II German army showed that many soldiers stubbornly continued fighting despite certain defeat. They had an intrinsic commitment to support their comrades and to do their duty to country. More rational people would have long given up and surrendered to save their lives.[15]

Extrinsic and intrinsic motivations appear in politics. When politicians promise rewards to voters and such rewards are not forthcoming, voter support may ebb. Jack Kennedy, on the other hand, inspired people, saying they should not ask what their country could do for them, but rather, what they could do for their country. His three years in office yielded little significant legislation to reward supporters. However, many continued to have a strong commitment to him and his dreams. Many young people joined the Peace Corps because Kennedy inspired an intrinsic commitment to help others, not because of promised extrinsic rewards.

Problems with Extrinsic Motivations

The problematic nature of extrinsic motivation appears in the high failure rate of dieting to lose weight. People often diet to get extrinsic rewards. They want to look better or please a mate. They join clubs where they get praise for losing weight.

Dieters have difficulties when external rewards fail. Perhaps the person they are dating, and wish to please, decides to leave. People may have to quit a dieting club because of the cost. Even the self-control needed for dieting has limits. It is like a muscle. At day's end or under stress, ego-strength may weaken. It becomes harder to keep the goals in mind.[16, 17]

Besides, going to a weight-loss clinic or following a prescribed diet often leads to a feeling that success came from the clinic's program or from the imposed diet. By accepting external controls, dieters give up personal initiative. They become pawns to the external rewards. They are less able to diet on their own, to make autonomous choices, than if they never had accepted such external controls.

It seems successful weight loss requires individual autonomy to value healthy eating as an internal reward, an intrinsic end. Value has to be attached to some behaviors that will aid in weight loss, perhaps feeling vigorous when slightly hungry. Dieters need to attach a sense of meaning to such feelings and behaviors. People who freely choose to diet will also feel a sense of competence or efficacy that helps them continue. Success requires intrinsic motivation, feeling good about the process, not simply the ends.

Research has confirmed that increasing self-determination and intrinsic motivation can yield better self-regulation in eating and in exercising, resulting in successful weight control.[18] The same principles have led to success by diabetics in achieving glycemic control and by people giving up their smoking.[19-21]

Intrinsic motivation is similarly involved when athletes attach meaning to the process rather than focusing only on goals. If they suffer a setback or failure, they can more readily shrug it off and continue working. They can feel, "I tried my best," or perhaps, "it was a learning experience." Einstein similarly stressed the process, saying, "The value of achievement lies in the achieving."[10,p413]

Adding extrinsic rewards to existing intrinsic motivation may harm rather than help. Money was given to students who enjoyed doing their homework—they already had intrinsic motivation. When the students no longer got money, the extrinsic motivation evaporated, but so did their intrinsic motivation. The students stopped being enthusiastic about doing the homework that they had once enjoyed. In a sense, the students started with an internal focus for satisfaction, but the added extrinsic rewards shifted their focus externally. It then became hard for them to regain the earlier internal focus.[22]

The use of external rewards can be seductive because they seem so simple to apply. People wanting to donate blood were also promised rewards, but they then became less likely to participate. Studies have also found that applying extrinsic rewards to creative tasks that are already intrinsically rewarding will reduce imagination.[23, 24]

Interviews at an amusement park in Vienna showed the harmful effect of extrinsic rewards on intrinsic commitments. People who were enjoying themselves were less able to name a source of personal meaningfulness, compared to people elsewhere not enjoying themselves. Entertainment with its extrinsic rewards may distract from a meaningful, intrinsically committed life that contributes to one's values.[25, 26]

Meaning and integration with personal values are important for paraplegics and others with similar serious losses. If they see their accident as meaningful, perhaps part of their love of auto racing or hang gliding, they can more readily accept their problems. It may be an opportunity to help others with similar afflictions, a chance to "make lemonade out of lemons."

There is much research showing that intrinsic motivation enhances happiness. People who are married, have strong religious beliefs (including atheism), and like their work (whether white or blue collar) are on the average happier. Extrinsic motivation can have the opposite effect, increasing depression and anxiety and decreasing psychological well-being.[27-30]

Needed Ingredients for Intrinsic Commitments

To develop intrinsic commitment, an individual needs autonomy, the freedom of choice and action. Then there is a feeling of personal responsibility for the results, including even results that are negative. Secondly, positive feedback from actions, feeling efficacy or competence, adds to intrinsic motivation. Finally, warm, positive relationships with others can foster intrinsic motivation.

All three, self-determination, efficacy, and relatedness, seem part of the human instinct to grow. Thus, everyone has the potential to develop intrinsic motivation.[31, 32]

Research shows the third factor, a sense of relatedness, is important starting early in life. Infants with a secure attachment to a parent are more likely to explore and engage with the environment. In contrast, children will develop less intrinsic motivation if an adult bystander ignores them or doesn't respond to attempts at interaction. Students who see their schoolteachers as cold and uncaring will naturally develop less intrinsic motivation than those who view their teachers as caring. Similarly, a study of workers with satisfactory interpersonal relations, as well as autonomy and competence, showed them to be productive employees.[32, 33]

Fostering intrinsic motivation in children requires some autonomy, not just the application of rewards and punishments by parents and teachers. If children come to enjoy the activities they discover and develop, they will then will seek out and conquer new challenges suited to their abilities. This is an important part of Montessori education.

Einstein's mother had him navigate a busy Munich street by himself when he was only four years old. The successful completion of this and other similarly difficult assignments on his own probably contributed to his lifelong intrinsic motivation to confront challenges despite many failures along the way. He would later say, "One should not pursue goals that are easily achieved. One must develop an instinct for what one can just barely achieve through one's greatest efforts."[10,p360]

Some parents may feel that they can mold their children simply with extrinsic rewards. This works in the short run but can be harmful if the children continue to have an external focus, looking for satisfactions only in the environment. While some external rewards and rules are necessary at first in parenting, eventually there should be internalization and integration of the rewarded habits, a "taking in" to be part of the self. Eventual freedom of choice is important so behaviors at first regulated by parents and teachers can become freely pursued.

With autonomy, children will eventually set their own goals and aspirations to excellence. For example, introducing choice in thinking about future careers can help students grasp the importance of doing relevant homework. In contrast, the more a student sees parents and teachers as controlling and applying pressure, the less the internalization or integration. As mentioned above, the more relatedness or interest and warmth from authorities, the more the behaviors will be internalized.

Developing Intrinsic Motivation

Forming an intrinsic commitment is often not fully rational. There is usually no time to consider all results from possible actions, so people act mostly on incomplete knowledge or partial rationality. People often look for a possibility that seems satisfactory, rather than optimal. For example, one may approach an attractive person just to see "where it goes."[34]

As a result, unforeseen things may appear, and there will be psychological pressure to make these meaningful. The results will be gradually assimilated into an intrinsic commitment, or perhaps not. Values can emerge or change with further actions. It is action, not a particular decision, that can foster a commitment.

Negative outcomes are often assimilated into an intrinsic commitment. There may be initial ambivalence and then a subsequent shift in meaning that eliminates ambivalence. People want to be consistent, to keep their integrity, and to preserve important beliefs and values. This often creates a sense of purpose that is beyond consideration of simple pros and cons.[30]

Thus, a lover may feel committed to someone, not despite some negative features, but eventually because of them. They come to be part of the beloved. They have intrinsic value and have meaning. One comes to see something as ideal that is actually not ideal.

In contrast, someone may find another person attractive because of extrinsic rewards, perhaps generosity and wealth. However, learning such beliefs are mistaken may give pause to the relationship. Negative results damage extrinsic motivations.

Research has shown the paradoxical strengthening of intrinsic motivation if negative results are assimilated. A classic study found that severe hazing increased commitment to a fraternity.[35] Another study followed cult members who had taken shelter in the belief the world would end. When they later emerged and accepted the negative results, their beliefs became stronger. They then sought new converts.[36]

In Chapter 2, we will see how members of a nineteenth-century German religious commune, Harmony, had to make many sacrifices such as renouncing assets and giving up sex. Paradoxically, Harmony lasted much longer than the many American communes that imposed fewer costs.[37]

Escalation in Intrinsic Commitments

Intrinsic commitments often intensify over time. When intrinsic motives are satisfied, there will be increased feelings of competence and self-determination that in turn heighten the intrinsic motivation. One is ready to continue acting in a self-determined way to find still more challenges.

Escalation occurred in the notorious prison simulation at Stanford University. Students in the role of guards at first felt they weren't real guards. However, as their early actions made the prisoners uncomfortable, guards embraced this as necessary and their feelings of competence increased. They eventually became quite cruel.[38]

Intrinsic actions often become harder and harder to break. This occurred in an experiment that had subjects give increasing electric shocks to foster learning by another person, a stooge. The experimenter didn't ask the people giving the shocks if they wanted to stop. Many of them became so involved that they kept increasing the shocks until the experimenter halted the study—at 450 deadly volts![39]

As commitments get stronger, less rewarding matters may become even less important because they can interfere with the intrinsic, often transcendent, commitments. Charles Darwin admitted that he lost his pleasure in poetry as he became more deeply involved in his theories of evolution.

Both intrinsic and extrinsic motivations can be frustrated and fail. However, such failure can be more readily accepted if the actions are freely chosen and the focus is on the process and the feelings of efficacy in choosing, instead of a focus on some external reward.

Thus trial and error behavior motivated by curiosity is a hallmark of intrinsic motivation. Einstein (Chapter 4) asserted, "the main source of all technological achievements is the divine curiosity and playful drive of the tinkering and thoughtful researcher."[10,p381] Bertolt Brecht (Chapter 6) continually tinkered with his plays—the staging, dialogues, lighting, and costumes. He wanted to see if anything better worked. Because he freely chose what to try, he wasn't frustrated by failure. Einstein similarly continued to use different initiatives over three decades in his unsuccessful attempt to combine gravity and quantum mechanics.

From Extrinsic Motivation to Meaningfulness

In Chapter 8, we will meet some postwar immigrants who tried various strategies to make life more meaningful. Rudy Geiger started with strong extrinsic motivation to become a millionaire, but he was forced to reconsider. Eva Bromke became committed to Catholicism as a child but then was frustrated by many of its tenets, such as the Immaculate Conception. She had a major struggle to deal with this. We will also meet young people who had a strong commitment to fight for Germany, a commitment developed from their membership in the Hitler Youth. Finally, we will see how Heinz and Ellen Biesdorf acquired a commitment to efficient and frugal living, a common commitment in Swabia. They remained single-minded in this even after becoming wealthy.

These people learned common cultural norms, as most children do. Parents, schools, and churches reward proper behavior and punish improper behavior. But will such habitual behaviors continue after the childhood rewards are gone? How can intrinsic motivation be developed so individuals do not need external supports? The stories here provide some answers.

1: CONRAD WEISER, PEACEMAKER

Swabian Wit: Two people on a train were quarreling, one stubbornly wanted the window open because he felt he was suffocating, the other stubbornly wanted it down, fearing an infection. The other passengers couldn't stand the strife and asked the conductor to lay down the law. He responded: "First we will open the window so the one dies of illness, and then we will close it so the other suffocates. Then we will have our peace."

Stubborn people can be selfish and thoughtless. However, sometimes stubbornness is a commitment to do good, as in Conrad Weiser, who doggedly fostered peace with the Native Americans. He was the most important interpreter of Indian culture in the eighteenth-century English colonies in North America.[40, 41]

Why They Left

Weisers had lived for many generations in the small village of Grossaspach, thirty miles northeast of Stuttgart. Conrad's father, John Weiser, was a baker and well-off farmer. In 1707, French troops launched one of their many invasions into the area. The winter of 1708-09 was the most severe in a century, ruining the vineyards. A final blow for the Weisers was the death of John's wife, forty-two years old and pregnant with her fifteenth child.

John Weiser was forty-five, but he bravely took most of his children and joined the first major wave of Germans to settle in eighteenth-century America. Many were called Palatines, after the Palatinate located on the middle Rhine. Some came from further upstream, like the Weisers and other Swabians.

They were mostly Protestants who were afraid that Catholic France would annex their territory. They feared the fate of the Protestant Huguenots who had been abused and driven out of France. Queen Anne of England, a Protestant, offered asylum to the Germans. Upwards of 10,000 Palatine refugees headed to London in the summer of 1709.

The English realized they would have to support these destitute people, so they quickly came up with ways to disperse them. Some were sent to colonize Ireland as well as the Carolinas in America.[42] Some 3,000, including the Weisers, were sent to the English Crown colony of New York to provide cheap labor producing tar and pitch for wooden ships. Once debts were paid off through their labor, they would get forty acres of farmland. The Indians on New York's frontier had given the land to Queen Anne to aid fellow Christians.

The Palatines were packed like cattle onto ships for the ocean crossing. Almost 500, including many young children, died during the six-month voyage to New York City. There, to reduce the Crown's expenses, the older children were contracted as workers for English colonists. The adults and younger children, including John Weiser's thirteen-year-old son, Conrad, were settled on Livingston Manor, about 100 miles up the Hudson River near present-day Rhinebeck.[43]

A Raw Deal

The Palatines found no land suitable for raising crops at their new settlement, so they had to buy food at inflated prices from Robert Livingston. He was a large landowner with an unsavory reputation including financial dealings with the pirate Captain Kidd. The tar production went badly through no fault of the Palatines. They were not paid and didn't get the land promised them. In protest, they went on strike. Soldiers put down the rebellion.

The English often mistreated immigrants to the colonies in this way. Later in the eighteenth century, for example, many Swabians were tricked into settling at Waldoboro, Maine. They didn't get what was promised by British General Samuel Waldo, and many eventually had to buy their land two or three times over. The smarter ones left in disgust and settled among other Germans in the Moravian town of Salem, North Carolina.

In the fall of 1712, the governor of New York announced there was no more money for the tar project. The Palatines would have to hire themselves out locally or get permission to go elsewhere in New York Colony.

John Weiser led a group of Palatines from Livingston Manor to settle along Schoharie Creek, which flowed north out of the Catskill Mountains into the Mohawk River. This was land the Indians had promised Queen Anne for Christian settlers. The destitute pioneers arrived with just the clothes on their backs. They fashioned crude tools and set to work. During the first year, Palatines in Schenectady and local Indians helped the settlers with food. Soon there were seven thriving villages, including one named Weisersdorf.

Living with the Mohawks

The next year, 1713, John Weiser allowed his sixteen-year-old son Conrad to live with the nearby Mohawk Indians. A missionary to the Indians apparently fostered the invitation. Conrad's life at home had not been good. "My father was influenced by my stepmother to be hard on me. I had no friend and suffered from hunger and cold. I often thought I would run away, but I didn't know where."[44,p135]

After arriving at the Mohawk encampment, Weiser became close to his foster father, the chief, and his family. However, during his stay of eight months Weiser sometimes had to hide from violent drunken Indians. In addition, he suffered from the cold, being "but badly clothed." He was also hungry, since the Indians didn't have much food towards spring. [40,p24]

Weiser's struggles were probably significant in strengthening his feelings of friendship with the Indians. Strong positive commitments can come from combinations of positive and negative events. The negatives are reframed to energize and enhance the positive.

The Mohawks were the easternmost tribe in the powerful Five Nations Confederacy, also known as the Iroquois. The Mohawk home territory was along the Hudson and Mohawk rivers in what is today eastern New York State. West from the Mohawks lived the Oneidas, the Onondagas (the keepers of the confederacy's council fire), the Cayugas, and, towards Lake Erie, the Senecas.

According to legend, Hiawatha, an Onondaga driven from his nation, came to stay with the Mohawks and convinced them to form a league of Iroquois Indians. Using an organization of clans with ties of intermarriage, the five tribes bound themselves together. They were later joined by a sixth tribe, the Tuscaroras, and became even more powerful.

The Iroquois had a deep sense of brotherhood. They believed in a great spirit, the Holder of the Heavens, who expected that they care for one another. Conrad Weiser described this strong religious spirit in terms reflecting his own spirituality: "if by the word religion we understand the knitting of the soul to God, and the intimate relation to, and hunger after the highest Being arising therefrom, then we must certainly allow this apparently barbarous people a religion."[40,p21]

Conrad Weiser also appreciated that, like his own people in densely populated Swabia, the Native Americans mostly got along together. He later wrote: "One can be among them for thirty years and more and never once see two sober Indians fight or quarrel."[40,p21] The stereotype of the fierce Indian arose after the Whites began to drive them off their lands.

Another Raw Deal

The Schoharie settlers got along with the Mohawks but not with the rulers of New York Colony, who insisted the Palatines did not have permission to leave Livingston Manor nor buy lands from the Indians. The Indian lands supposedly belonged to a syndicate of seven merchants and land speculators, including Livingston. In 1715, the Governor ordered the Palatines to pay for their land or move on. The settlers refused, and when a sheriff arrived to arrest John Weiser, Palatine women threw the sheriff in the pigsties. He was then ridden on a rail through the settlements and beaten before being released.

Eventually John Weiser and several other leaders were caught, imprisoned, and forced to sign leases. Some settlers also reluctantly bought their farms from the authorities. Others moved to free government land in the wilds further north along the Mohawk River. Still others made a new home in more hospitable Pennsylvania Colony.

Conrad Weiser was in no hurry to leave. In 1720, he married a Palatine neighbor, Ann Eve, and then began to raise a family and develop his farm. In 1729, after several fruitless efforts to buy a larger tract for his family, the thirty-three-year-old Weiser moved to the Tulpehocken Valley in Pennsylvania, near present-day Reading, where his father-in-law was already established.

Weiser built a stone house (it still stands in Conrad Weiser State Park), settled down as a farmer, and took part in the community. He was selected as an elder in the local church, appointed to a committee to lay out a new road, and became Overseer of the Poor for the township. Weiser had many Germans as neighbors, since the fertile area south of Blue Mountain in southeast Pennsylvania attracted many immigrants in the early eighteenth century.[45]

Interpreter for Pennsylvania

Most importantly, Conrad Weiser began his remarkable role as Pennsylvania's link with the Iroquois. The local tribes and the white colonists in Pennsylvania were clashing. William Penn's policy of brotherly love in dealings with the Indians did not always prevail. Settlers, especially the often hot-headed Scotch-Irish, were pushing ever deeper into Indian lands. These settlers feared the Iroquois might come down from New York to defend their allied tribes against abuse from the Whites.

To prevent further trouble, the Provincial Secretary of Pennsylvania wanted an alliance with the Iroquois. He saw Weiser's knowledge of their language and culture as helpful. Instead of the Indian traders usually serving as interpreters, Shickellamy, an Oneida sent by the Iroquois as a negotiator, recommended Weiser. This resulted in a paid job of Provincial Interpreter for the colony. A chief of the Onondagas later told the English colonists that after the Indians adopted Weiser, they kept a part of him for themselves and gave the other part to the colonists.

In 1732, Shickellamy came from the Iroquois Council Fire in Onondaga to Pennsylvania with a large delegation of chiefs to act for the rest of the Six Nations. The talks lasted several weeks, with speeches and exchanges of wampum (shell beads) and other gifts. As was often the case, no written treaty resulted. But Weiser's work as interpreter and his friendship with Shickellamy were critical in forging ties between the Indian chiefs and the colonial leaders, including governor Thomas Penn and members of the governor's council. At the end, the chiefs asked that Shickellamy and Weiser manage future conferences.

Shortly after this conference, the Indians agreed to sell disputed lands in the Lebanon Valley to Pennsylvania (New York had previously sold them illegally to Pennsylvania). Weiser's home was there. His farm soon expanded to 200 and then 800 acres. He worked hard and invested his money first in horses and cattle, then in more real estate and a tannery. He became a respected business leader in the community.

Spiritual Life and Ephrata Cloister

After his arrival in Pennsylvania, Weiser had become a leader of the local church. Weiser was a student of the Bible and works of eminent Pietists and mystics, as was common in Swabia. He accepted the Pietistic idea that Christ is present within believers. More important than the Christ who died for one was the Christ who lived *in* one. Under Weiser's religious leadership, the congregation "sang, prayed, exhorted, admonished, and searched their souls and one another's."[40,p53]

This unified church included members of both established German Protestant churches, Lutheran and Reformed. However, the two groups eventually quarreled. They locked each other out, threw stones through the parsonage windows, and even put gunpowder in the firewood. This disgusted Weiser, and he left the church.

Weiser then spent six years deeply involved with the mystics at nearby Ephrata Cloister. These Germans were part of the Radical Pietist movement that sent several groups to America, including the Harmonists to be described in Chapter 2.[46] The leader at Ephrata, Conrad Beissel, stressed living a pure life, abstaining from sex, and engaging in a mystical union with God and the universe. Weiser was converted by charismatic Beissel, "whose face shone with an inner light and who seemed to walk in an aura of God's peace."[40,p58] Beissel had such personal magnetism that he once won a bride from her bridegroom at the marriage service.

Weiser still lived on his farm but often joined in the life of Ephrata monastery. He continued to father children but was otherwise a committed member of the ascetic community.

Eventually, however, Ephrata's mysticism became too extreme for Weiser. The community conjured ghosts, created mystical charms, and believed in occult practices and miracles. When Beissel had himself deified, Weiser's Swabian independence and sensible nature led him to join a faction that called Beissel "Brother," not "Father."[40,p104] To Weiser, Ephrata had become a cult. He distanced himself after one of the sisters at Ephrata was spurned by Beissel and she then accused him of impregnating her and later killing their child.

The Moravians

Weiser's spiritual search next led him to the Moravian Church, which followed the fifteenth-century teachings of John Hus. The Moravians had suffered much persecution in Europe and eventually found protection under Count Zinzendorf in Saxony. Groups of Moravians then left for America to set up missions and settlements at Bethlehem and Nazareth in Pennsylvania and elsewhere in the colonies. The Moravians had a vision of peace and union among all Christian churches. Weiser had worshiped with various faiths, even Roman Catholics and Jews, so he found the ecumenical Moravians attractive.

Zinzendorf came to America to lead missionary work. However, he did not view the Indians as equals. He denigrated their intelligence and said they lacked any signs of grace. He even complained about their reluctance to wear breeches and other "proper" clothing. His nobleman's views of the Indians contrasted sharply with Weiser's and those of most Germans in America.

Count Zinzendorf employed Weiser to take him on a trip to convert the Delawares and Shawnees. When Weiser returned after a short absence from Zinzendorf's missionary campsite, he found the count in trouble. The Indian guides and interpreters supplied by Weiser had abandoned the arrogant count. They told Weiser that Zinzendorf disdained them, pitching his own tent far from theirs and spending time reading books.

Zinzendorf felt Weiser had intentionally taken him to visit stubborn Indians. The count complained the natives showed no respect and even farted in his presence. Weiser later complained that Zinzendorf "never had his fingers properly rapped, instead he was always treated as the high-born count."[40,p242] Zinzendorf gave orders to the Indians, and they weren't used to taking orders from anyone.

Go-Between to the Iroquois

Weiser continued to be an interpreter and go-between for the Iroquois and the government of Pennsylvania. As part of this work, he took a six-week trip from his Tulpehocken home to the headquarters of the Six Nations at Onondaga. It was through the rugged "endless mountains" in northern Pennsylvania, still sparsely populated today.

At one point, Weiser was so exhausted and depressed that he sat down to die. His Indian companions noticed him missing, and they returned to urge him onward. As recorded by Weiser (and later published by Benjamin Franklin), Shickellamy encouraged him, saying that "evil days are better for us than good, for the first often warned us against sins, and washed them out, while the latter often enticed us to sin."[40,p89] It was a persuasive appeal to Weiser's ascetic nature.

Around this time, the French built a string of forts in western Pennsylvania, beyond the Allegheny Mountains. The Pennsylvania Assembly, controlled by pacifist Quakers, did not want to respond to the French threat.

Weiser tried to aid the Penns, the proprietors of Pennsylvania Colony, and their political party, by urging the German settlers to vote for people who would improve the colony's military defenses. But his efforts couldn't prevent the repeated election of the Quakers. Weiser also failed when he tried to run for office himself. Like so many other German immigrants, he was not adept at campaigning.

However, Weiser played a key role in a big conference at Lancaster in 1744. Weiser helped the Indians and English colonists in and around Pennsylvania better understand one another.

Virginians attended the Lancaster conference because the Indians were threatening war. Unfortunately, the Virginians typically acted superior to the "savages."[40,p189] Weiser warned the southerners on their first visit to the Lancaster Indian encampment not to laugh at their dress, or make any remarks on their behavior. A particular problem was that colonists had moved beyond the earlier agreed-upon border of the Blue Ridge Mountains and were settling in the Shenandoah Valley of Virginia. Fortunately for the Virginians, the Indians accepted payment for the stolen lands.

After two weeks of the conference, the Six Nations promised to side with the English against the French. Several months earlier,

France had declared war on England. If the Iroquois and their Indian allies had joined with the French, the English colonies would have become small coastal enclaves encircled by the French and Indians.

During the talks at Lancaster, the Iroquois chief, Canasatego, suggested the English colonies form a union like the Iroquois, a confederacy that would ensure strength. Weiser later delivered a transcript of the 1744 Lancaster Treaty talks to his good friend, Ben Franklin. After studying the report of the Iroquois political arrangement, Franklin would later propose a similar federal government. Separate British colonies, like the Indian tribes, would keep their sovereignty and constitutions, except where superseded at the federal level, analogous to the Council Fire at Onondaga. Aspects of the Iroquois political system were implemented first in the creation of the Continental Congress and later by the United States Constitution.

New York Colony

After his good work at Lancaster, Weiser accepted an invitation by Governor Clinton of New York Colony to look into their issues with the Indians. However, when Weiser honestly stated how his Indian friends resented Albany's policies, he was told his services were no longer needed. He was even forbidden to visit the Six Nations Council Fire in Onondaga, New York.

During an earlier visit to New York, an Onondaga chief told Weiser how he had visited the house where the Whites met every seventh day and reportedly discussed "good things." The chief did not know English, but he noted that a man dressed in black spoke angrily. If the Whites met so often to discuss "good things," the chief asked Weiser, why didn't they treat the Indians better. White travelers were welcomed in Indian homes, but an Indian traveling to Albany in New York is only asked, "Where is your Money?" If he has none, he is told, "Get out, you Indian Dog."[40,p226]

Trouble with the French

Following the rejection by New York Colony, Weiser turned his attention to the Ohio region. The tribes there—the Shawnee, Delaware, and Wyandot, plus young Iroquois looking for better hunting lands—had sent a delegation to Philadelphia to ask for support against the French. In 1748, Weiser journeyed over the Allegheny Mountains to meet with them in council along the Ohio River. He became convinced that strong action was needed to help the Indians stop the French from coming up the Ohio River into Pennsylvania. In 1751, the Shawnees sent word to Pennsylvania that they wanted aid. Nothing was done.

Finally in 1754, the Virginians under Colonel George Washington mounted a military expedition to the West. It was too late. They found nearly a thousand French already settled at strategic Fort Duquesne (Pittsburgh). In addition, Colonel Washington's inability, or reluctance, to understand the Indians led to their abandoning him. Without Indian support, the French defeated the Virginians at their hastily erected Fort Necessity. Washington had to surrender.

Word soon came that hundreds of Indians allied to the French were headed east to harass the settlers on Pennsylvania's frontier. Weiser helped gather some friendly Indian support for an English expeditionary force. But the English general, Edward Braddock, was no better than Colonel Washington. He had contempt for the Indians, so almost all of those recruited by Weiser left. Braddock sent away the rest, seeing them to be more trouble than they were worth. A small force of French and their Indian allies eventually routed the English, and General Braddock later died of his wounds.

Scalpings on the Frontier

When the few Indians still allied with the colonists repeatedly begged for soldiers and munitions, the Governor of Pennsylvania and the Quaker-controlled Assembly continued to disagree. What happened next was predictable. Encouraged by the French, Delaware Indians engaged in scalpings and massacres along the frontier. Still,

Philadelphia officials couldn't agree to send any help.

Weiser took action to protect his home and the Tulpehocken Valley. Like many German immigrants, he was not aggressive but ready to fight if necessary. (Later in January 1776 during the American Revolution, Weiser's son-in-law famously stood at a church pulpit, took off his pastor's clothes to reveal a Continental uniform underneath. He announced that there was a time for everything, a time to preach and a time to pray, but there was also a time to fight, and that time had come.)

Weiser got news of the first massacres by the Delawares at ten o'clock on a Sunday night, and he sent messengers ordering settlers to meet at his house by daybreak. He organized 200 men. But this did not halt the mounting confusion and terror elsewhere along the frontier. The settlers needed regular troops and forts.

One of the Iroquois chiefs came to Weiser and demanded to be taken to Philadelphia. He spoke to the Pennsylvania Assembly, insisting they act more strongly. They should act like men, not weak like women. But again, the Quakers dominating the Assembly refused to pass the Governor's bill appropriating money for defense.

Hatred of the Indians

As the attacks and scalpings by the Delawares continued, the settlers began to fear all Indians. They wanted a bounty on Indian scalps. However, Weiser kept the commitment to his brothers. He reported escorting his Indian friends, those he relied on as emissaries to the Six Nations, through the enraged settlers. Weiser "saw about 4 or 500 Men, and there was a loud Noise; I rode before, and in riding along the Road (and armed Men on both Sides of the Road) I heard some say, Why must we be killed by the Indians and we not kill them? Why are our Hands so tied?" Weiser got his friends to his house "where I treated them with a small Dram, and so parted in Love and Friendship."[40,p414]

Many settlers on the frontier wanted to burn down the houses of the few Quakers that lived there. A mob of mostly German settlers gathered in Philadelphia and demonstrated against the Quaker-dominated Assembly. Eventually a bill passed to issue money for powder, lead, muskets, and blankets for the frontier.

Gradually over the winter of 1755-56, Pennsylvania built up its army. Weiser, as a colonel, played a major role. But the scalpings, mutilations, and the destruction of houses, barns, and cattle continued.

The Pennsylvania government wanted a peace conference with the rampaging Delawares, but not all the colonists agreed. Quaker politicians led by Israel Pemberton worked to support the Delawares and their chief, Teedyuscung, against the party of the Penn proprietors. Pemberton apparently hoped the Quakers could escape blame for the colony's lack of preparation against the Delaware attacks.

Weiser Is Stripped of Power

The English had promoted New York's Indian agent, James Logan, to manager of Indian affairs for all the colonies, and forbade the proprietors of Pennsylvania to make any treaties with the Indians. Thus, Weiser no longer had an official position as Pennsylvania's ambassador to the Indians.

This was a serious blow to Weiser's friendship with the Indians. One of Weiser's Indian friends asked if he were more interested in his military commission than in their welfare. Weiser responded, "...don't charge me with such a thing, as that I take greater Delight in War than in civil Affairs. I am a man for peace, and if I had my wish there should be no war at all, at least not on this Side of the Great Waters."[40,p467]

However, Weiser continued to work with his brothers, the Indians, as an interpreter. He didn't let the loss of extrinsic rewards stop him, the removal of pay and status as ambassador and colonel. His intrinsic motivation was too strong.

As treaty negotiations dragged on, the self-righteous Pemberton acted smugly without knowledge of Indian culture, history, or the relationship between the Delawares and the Iroquois. Weiser knew Pennsylvania could not unilaterally grant the Delawares their wishes without considering the Iroquois. If Pennsylvania alienated the Iroquois, they would join the French and together wipe out the English colonies.

Weiser, now sixty and in failing health, was criticized by the rough frontiersmen for his Indian sympathies. Some said he was afraid of the Indians and questioned how he handled his military duties. After a letter from the Governor's secretary filled with biting gossip, petty complaints, and hypocritical accusations, Weiser resigned his battalion command. However, he could not give up his commitment as interpreter for his friends.

A Stubborn Triumph

In the treaty conference of 1758, Weiser single-handedly negotiated agreement between the Delawares and the Six Nations so the Indians as a whole could settle with the colonists. Though New York Colony's Logan had banned Weiser from any dealings with the Indians, Weiser stubbornly insisted that he would continue to interpret. Mostly by informal politicking in his room, he got the Delawares away from Pemberton's influence. Weiser's knowledge of Indian culture and mentality saved the conference from ending in disaster.

As a result, the Iroquois removed their allies, the Delawares, from the side of the French. This ended the war in the west and in Pennsylvania. On the other fronts of the French and Indian War, Quebec fell and then Montreal. The English colonists triumphed over the French.

Had Weiser lived longer, he might have continued to play an important role in mediating between his Indian friends and the colonists. But with the defeat of the French, the English colonists no longer felt the Indians were important. Soon, the Indians to the west of the thirteen colonies would unite against heedless invasions by the Whites. Because of his ill health, Weiser could not forestall the violence of Pontiac's Conspiracy and the tragic slaughter of the Indians as settlers moved westward. On July 13, 1760, at the age of sixty-three, he died.

After his death Weiser was eulogized by the Indians as "a great man, and one-half a Seven Nation Indian, and one-half an Englishman."[40,p71]

2: COMMUNES

A frugal Swabian peasant went shopping for socks for her husband. The clerk asked how many pairs she wanted. She retorted angrily, "Are you accusing me of having more than one husband?"

Some Swabian immigrants combined great frugality, religious fervor, and communism in a few successful nineteenth century American communities. The notoriously individualistic Swabians didn't find communism an attractive political doctrine. They were simply following *Acts 4:32* by having all things in common. These religious communes stubbornly survived for up to a century—even after the expected Second Coming of Christ didn't happen.[47, 48]

Rapp, a Charismatic Pietist

Johann Georg Rapp was born in 1757 to a peasant family in Iptingen. When told as a youth to harvest grass for the cattle, he convinced schoolmates to do the job while he preached from a tree. It was a sign of things to come.

Rapp attended meetings of the Pietists in homes as well as services of his church. The Pietists believed there was a lack of spiritual feeling in the state-run Lutheran church, but Rapp felt that even Pietist meetings fell short of his spiritual ideals. He accused one Pietist leader of being hypocritical and covetous.

Rapp began preaching publicly, and his charismatic personality attracted many followers. The Lutheran authorities wanted explanations. Rapp asserted that he had rejected church and communion because he had personally found Christ. Since he used only the Bible, a book of Luther's writings, and instruction by the Holy Spirit, Rapp claimed he was not a heretic. Rapp was on a mission, saying "I am a prophet and am called to be one."[49,p1]

Rapp described his decision to separate from the church: "I felt so small that I did not care to open my eyes. It was then that I saw that salvation is in Christ Jesus alone, and also saw how necessary it was to be in a still place and to hearken to one's heart. Accordingly, I was convinced that the soul's very own life, as small as it may be, must die so that Jesus may become its resurrection and life. Since that time also I have not come to church again, because my Jesus shone into my heart brightly as the word of life, and because I needed nothing else since I had found Jesus."[50,p18]

Asked why he did not follow the Biblical command to obey his teachers, Rapp said there were no longer any true teachers. The Apostles had not needed university degrees like the pastors employed by the state-run Lutheran church. Rapp said the wealthy educated clergy were no longer Christ's servants but merely in it for the money. In this, Rapp expressed many peasants' long-standing dislike of their masters.

Rapp also thought that people going to the Lutheran churches weren't committed Christians: They confessed with their lips but lived as pagans. They prayed but acted contrary to God's rules.

Separatists

State officials spent four years looking into Rapp and his followers, who came to be known as Separatists because they wanted their own preachers and congregations. The authorities recommended kind instruction so they could renounce their errors. If they did not, they should be expelled from the realm for disobeying Württemberg's religious laws.

When leaders of the Separatists were later imprisoned, followers demanded that they be imprisoned, too. When the leaders were released, citizens loyal to the Lutheran Church were angry and appealed to higher authorities including the Duke of Württemberg. Rapp refused to recant, saying it would be a great sin to deviate from the Spirit's direction.

As many others became sympathetic to the Separatists, the authorities realized that Württemberg could not afford to lose so many energetic and ambitious people. Hoping reason would prevail, the government issued a list of minimum requirements for the Separatists. They had to pay the compulsory Lutheran religious taxes that everyone else paid. Meetings at night or during regular church services were forbidden. Meetings at other times were okay. In addition, people who were not local were barred from Separatist meetings. People came anyway, from up to thirty and forty miles away, extraordinary distances to travel in those times. Regular Lutheran services were neglected.

Rapp was a mystic influenced by Jacob Böhme's Gnosticism, which postulated the dualism of nature and the need to free the soul trapped in a material being. Germans inspired by Böhme later founded the Amana communes in Iowa.

Rapp correlated the prophecies of the Book of Revelation with events in Europe. Napoleon had brought major changes to Swabia, sweeping away traditional laws and governmental units. Was this a prelude to the Second Coming and the resurrection of the souls of the dead? Unlike the Lutheran pastors, Rapp thought so, and his followers grew to more than 10,000.

The Separatists eventually gave a written account of their beliefs to the Württemberg legislature: (1) a church based on inner conviction and separate from the organized church; (2) baptism only for those moved by God; (3) communion only after unity of feeling in the assembled group; (4) independent schools for Separatist children; (5) no confirmation ceremonies since they were not based on the feelings of the heart; (6) no oaths to the government; instead, obligations should come from inner feelings of duty and love, with the payment of taxes or money in place of military service.

Württemberg officials felt threatened. Some may have feared a repetition of the Peasant War of 1525. Religious fervor might again demand democracy, including the election of pastors and rule by the Bible, not the government.

Abandoning their previous leniency, religious officials now recognized popular hopes for the Second Coming of Christ but insisted that government-appointed clergy were in charge of instruction. If the Separatists persisted in disturbing order by holding meetings of more than fifteen people, all would be imprisoned and their leaders severely punished.

Faced with these threats, Rapp and some followers left in 1803 to find land in America where they could practice their religion and prepare for Judgment Day. Rapp was following an interpretation of *Revelation 12:6* that God's Kingdom would be established in a wilderness while Europe would be destroyed.[51] A century earlier, Swabian Johann Albrecht Bengel had prophesied that Judgment Day would come in 1836.

Next summer, large groups of Separatists emigrated and settled in parts of Pennsylvania and eastern Ohio while Rapp looked for affordable land for a communal Separatist settlement. Rapp tried to buy 40,000 acres in Columbiana County, Ohio, but didn't have the money. He met with President Thomas Jefferson for help but only Congress could arrange such a deal. Congress balked. Instead, Rapp bought a small tract of land north of Pittsburgh.

In forming Rapp's Harmony Society there, the 500 members signed a simple set of articles of association, transferred their wealth to Rapp and to the society, agreed to obey society laws and chosen leaders, and, if they wished to leave, not to ask payment for services rendered. Rapp and the Harmony Society agreed to accept members, instruct them in religion, provide their subsistence, and, if they were to withdraw, refund their contributions. By surrendering their wealth, members made the first of many sacrifices that would develop strong commitment enabling the commune to survive indefinitely.

As more Separatists arrived, the amount of land proved inadequate. The commune also needed a land buffer from the secular influences of the outside world. In addition, the climate seemed unsuitable for cultivating the grapes they had grown in Swabia. Planned manufacturing was also behind schedule. Financial worries and internal quarrels erupted. Some left the Society and filed lawsuits.

In 1806, Rapp again went to Washington to ask for federal help in getting land, especially in a warmer climate where they could have vineyards. The Senate passed a bill, but the House refused, in part because whiskey was the American drink, not wine. [52,p18]

Sacrifices

Rapp became firmer in consecrating members to a total, pure Christianity. This would make them ready for the Final Days. At Rapp's urging, the Society decided to adopt celibacy for all, hoping to ensure the Harmonists would have the highest commitment possible. However, some felt doing without tobacco was an even greater sacrifice than giving up sex. Celibacy would reduce commitment to a partner in place of the commune. Interestingly enough, some other long-lived communes such as Oneida ensured similar high organizational commitment by requiring free love instead of celibacy.[37]

Research has revealed that the sacrifices by members fostered commune survival. Many sacrifices were common in nine communes lasting from 33 to 184 years and less common in twenty one communes that lasted only 6 months to 15 years.[37]

The guiding principle of Harmony was: "We endure and suffer, Labour & Toil, sow and reap with and for each other." Further, "The spirit of our age abhors the useless works of vanity and of self-interest."[53,pp49,68]

Sacrifices were strictly enforced. No one could read a book or newspaper or talk to neighbors because that would reduce the total commitment to work.[54] Other distractions were eliminated: English was not allowed, members could not talk with neighbors, and visiting by outsiders, even Germans, was also strictly controlled. Sacrifices by commune members were also fostered by Rapp's radical Pietism that stressed "passivity before God" and "annihilation of self-will." This all helped ensure that commitment to work and the commune had the highest priority. [49,p9]

Harmony stressed work because Rapp believed that Christ would need money for his coming thousand-year reign on earth.[49] The Harmonists also built on the Swabian tradition of self-reliance. God helps those who help themselves, so Harmony tried to be economically independent of the outside world. They built a gristmill, an oil mill, a tannery, a dyer's shop, a warehouse, a saw mill, a brewery, and an inn, all before they built their church. As mystics, they felt God's spirit was within each person. They could worship outdoors or wherever the Spirit moved them.

Rapp's firm leadership helped the Society to endure frontier hardships. They frugally plowed earnings back into the commune rather than spending on consumption. To sell products to others, they also built mills for wool and hemp, a second gristmill, a second warehouse, a soap and candle works, and carpenter and machine shops. When they later moved to Indiana, they built factories for bricks, potash, rope, nails, and other products.

The move to Indiana in 1814 was driven in part by increasingly jealous neighbors in Pennsylvania. There was also anger over Harmony's refusal of a military call-up for the War of 1812.

The new location on the Wabash River had waterpower and a connection to the Ohio River. Grapevines grew well. Because whiskey, not wine, was popular on the frontier, Harmony built a distillery and a nearby whiskey sales shop. Members couldn't drink the whiskey, only their strictly rationed beer and wine.[49]

The Indiana settlement prospered. New members came from Swabia, and both farming and manufacturing expanded. The Harmonists also made great spiritual progress. The Society was productive in everything except children.[55]

Return to Pennsylvania

After ten years in Indiana, Society leaders realized they needed to be closer to large markets for their many manufactured goods. Neighbors also envied the commune, a wealthy enterprise amid rough frontier life. The Americans did not understand that seemingly ignorant Germans had earned their success through frugality and hard work. Some did not appreciate the educated culture of the Society. Harmony's library, after all, was larger than Indiana's state library. Only thirty-nine of the 500 members were illiterate, a much smaller percentage than among the surrounding pioneers.

However, the Harmonists were not intellectuals. They believed in a harmonious development of body, mind, spirit, and soul. Like most other immigrants, they were people of action who distained those who merely thought without acting. When a professor applied for membership, Rapp told him to hoe potatoes to counteract his "abnormally developed mind."[50,p252]

Like most Swabians, the Harmonists valued music. The Society fostered music instruction, and concerts were frequent. However, members could not become professional musicians. Community singing, like other group rituals, helped foster commitment to the commune as a whole.

Commitment was no doubt strengthened by the hardships endured, first the difficulties of settling on the frontier (twice), as well as sickness and deaths, especially from the malaria along the Wabash River. Surviving the hardships helped members feel that belonging to their commune was "worth it."

In addition, Harmony and other successful communes used mortification to erase the "sin of pride," reducing egoism in favor of greater commitment to the commune. The emphasis on humility and self-criticism could be unpleasant for individuals. However, when endured and then assimilated, it made for strong commitment and commune longevity.[37]

Egalitarianism ruled in all matters, even after death. To prevent some of the departed from having more mourners than others, the commune tried to limit those present at final rites to the first twenty four to appear in the procession to the cemetery. When not enough mourners appeared, the Society conscripted more.

In 1824, the 700 Harmonists moved back to Pennsylvania. They sold their land and town to Robert Owen. His utopian society lasted but two years. [52]

The Harmonists' new settlement on the Ohio River just west of Pittsburgh was better for marketing and had more agreeable neighbors. The Harmonists called their new settlement Economy. Rapp expected to live out his years there while waiting for the Second Coming. In short order, the commune surpassed its productivity in Indiana. Despite the great commitment to work, the visiting Duke of Saxe-Weimar reported that members were happy. In 1826, he noted that factories and workshops were heated in winter by pipes connected to the steam engine. All, especially women, had "very healthy complexions" and greeted Rapp "with warm-hearted friendliness." There were fresh flowers on the machines.[56,p93]

However, there were problems. Though Rapp had mandated celibacy to heighten spiritual development in preparation for the Second Coming, some members could not restrain themselves. They were forced to leave if they had children.

Others left voluntarily to marry, but they had been so committed to Rapp's teaching of celibacy that they often felt guilty about sex. One wife reported after deserting her new husband: "We have no luck together anyway because the curse rests upon us."[50,p422]

Rapp himself was attracted to a twenty-year-old woman, Hildegard Mutschler, his laboratory assistant. Conrad Feucht, the Harmonists' physician, was also smitten with her, so Rapp forced him to leave. Mutschler then ran off to be with Feucht. Rapp asked them both to return as husband and wife, but other Society members found it hard to stomach such favoritism. The Feuchts returned, lived separately but had three children. Rapp let them stay.

Rapp's favoritism to Mutschler led some Society members to leave and caused a falling-out with his adopted son, Frederick, who was second-in-command. A visitor at the time recorded Rapp scolding his congregation, especially the children born in America, for their disobedient ways.

The Ambassador of God

As the Harmonists became increasingly tense over how to prepare for the predicted Second Coming in 1836, an astounding letter arrived from a Count Leon. The letter showed familiarity with the history of the Society and its innermost mystical secrets. The Count wrote that the Second Coming was at hand and the Lord had anointed him to act in His name.

Two years later, the Count and his followers triumphantly arrived. The aristocrat acted superior to the peasant-born Rapp, and some Harmonists encouraged the count to take his rightful place as leader. They urged him to allow sexual relations and greater material pleasures. Others who had left Harmony and now lived in poverty asked Count Leon to help them get their rightful share of Society money.

Rapp's son, Frederick, the Society's business manager, presented a bill to the count for his stay. However, a third of the Harmonists voted for Count Leon as their new leader. Separate groups formed, each claiming to represent the community. Each filed for debts and damages from the other. Some minor physical violence even occurred with Irish immigrants hired to do the fighting.

The two opposing groups finally reached a financial settlement in March 1832. Count Leon and his supporters left to build a new community nearby, but it was not successful. The count was less strict, so some of his more individualistic followers set out independently.

Conflict between the two communes arose over the payouts to be made by the Harmonists, and the members who had left sued for breach of contract. However, because the Harmony Society was never incorporated, lawsuits could be filed only against Rapp personally by those who were still members. Ex-members asked for reinstatement but were rebuffed.

Count Leon then ran out of money after his alchemy to create gold failed. Even a reinforcement of some 400 Bavarian Catholics didn't help his struggling commune.

It became known the count was the illegitimate son of a seamstress and a German nobleman. He had squandered his great inherited wealth in various escapades in Germany.

Rapp sued the count, who fled a day before his scheduled court appearance, taking his original following and a few others. They settled in Louisiana, but many eventually fell ill and died there.

The dedicated, hardworking Harmonists overcame the financial setback from Count Leon, but in 1834 Frederick, Rapp's adopted son, died. He had handled the increasingly important businesses and external affairs of the commune. Manufacturing and trade had replaced farming as the economic foundation of the Society. In his late seventies, Rapp delegated much of Frederick's work to two men: Jacob Henrici who had grown up in the Society, and R. L. Baker, a Palatine schoolteacher who had arrived during the third settlement.

Though some claimed the Harmonists were slaves of Rapp, many did leave the commune freely. The community accepted that some weren't able or willing to follow the rules and remain. By allowing free choice, those who stayed became even more committed to the commune.

Membership Is Closed

With its great wealth attracting many strangers, the Society halted new membership. Those who had remained loyal, working hard and denying themselves, felt that undisciplined outsiders could not match their spiritual development. They had not made sacrifices, so how could they have the same fervent commitment to the commune?

Rapp sealed the treasury from claims by any future defectors. The 391 existing members agreed the Society forever owned their contributions on joining. This cemented the commitment of all remaining members.

Finally, Rapp made the commune free of outside economic influences by converting its half million dollars of paper money to a hidden hoard of gold and silver. Then they burned the account book that recorded this wealth. The Harmonists would follow the Swabian ideal of "having, not showing." They would continue to work hard as long as they could or until the Second Coming of Christ.

A letter written by Rapp three years before his death described the Society's secret of success: mostly frugality and self-denial. Unnecessary things such as tobacco were prohibited. Their simple clothing and food came almost completely from within the commune. Nothing was spent on imported fashion and luxury items. They were still influenced by the strict culture fostered by Swabia's 200 years of religious courts established during the Protestant Reformation.[5]

For some, the restrictions were undoubtedly sacrifices. On the other hand, when intrinsic motivation is very strong extrinsic rewards no longer matter. We will see that others with strong intrinsic motivation, people like Albert Einstein and Bertolt Brecht, lived simple, ascetic lives. Extrinsic rewards are no longer important. In fact, extrinsic rewards get in the way of dedicated intrinsic motivation.

The Harmonists were conscientious in everything they did. They practiced intensive agriculture where nothing was wasted. Fields too hilly to cultivate were planted with mulberry trees, marshes were used to grow willows for basket making, and a useless ravine served as a pigpen. To this day, grapevines can be seen trained up the walls of the Harmonists' buildings, freeing-up land for crops instead of a vineyard.

Until his death in 1847 at age eighty-nine, Rapp preached the Second Coming was still close. The two succeeding trustees continued to carry on in Rapp's spirit. Henrici became the Harmonists' religious leader and Baker their secular leader. The latter stressed that members needed to continue exercising self-denial.

Celibacy continued to be the rule despite a declining population. The only young people were the offspring of the Feuchts. Most Society members were the adult children of the early colonists. Membership declined from 288 in 1847 to 146 in 1866. Gradually the society gave up some manufacturing, such as the labor-intensive silk industry. Clothing was purchased from outside, if possible from the German commune in Amana, Iowa. Half the farmland was rented to sharecroppers, and outsiders were hired as laborers for other enterprises.

Enormous Wealth

Investments became very important. Harmony's stock portfolio yielded more than the work of members. Harmony owned one-third of the nearby town of Beaver Falls, including a knife and tool factory, two stove factories, two china factories, and two wood planing mills. The knife factory alone employed 300 workers.[57] The Society paid $650,000 for a railroad line from Pittsburgh north to Youngstown, Ohio, connecting the east-west lines of the New York Central and the Erie railroads. The highly successful new line was later sold to Cornelius Vanderbilt. Along with other railroad ventures, the Society also invested in coal mining and became a significant player in the newly discovered oil fields of northwestern Pennsylvania.

Despite their wealth, Society members continued to live according to the New Testament, which asserted that followers of Christ should treat their goods as though God owned them. They continued to work hard and became more set in their ways. They were prepared to wait as long as it would take for the Second Coming to happen.

Ex-members continued to sue the Society, and the Harmonists hired the best lawyers and stubbornly fought back, all the way to the Supreme Court. As a matter of principle, the Society spent more money on these cases than the disputed sums.

We often think of frugal people as selfish. However, at Harmony religion fostered not only tightfistedness but also dedication to worthy causes. The Harmonists charitably fed and sheltered homeless people. At certain times, such as the Fourth of July, they invited neighbors to share in their plenty.

Henrici also supported other religious enterprises. He sent money to Iceland for the poor in a religious colony and to the Temple Society and a colony of Swabians in the Holy Land. Henrici also donated to the Hutterites at their Russian settlements. The Hutterites shared many beliefs with the Harmonists, except for celibacy. When they left Russia, Henrici even tried unsuccessfully to have them settle on some of Harmony's underutilized land. Henrici also helped the Shakers and the German communes of Zoar in Ohio, Amana in Iowa, and *Ora et Labora* in Michigan.

Swindler Duss

By 1890, Harmony's membership had dwindled to a few very old people. Several individuals previously connected to the Society arrived to join, including John Duss and his mother and wife, as well as the two sons of the Feuchts.[58]

Duss had been two years old when his widowed mother brought him to Harmony in 1862. They had worked there intermittently over the years but were never allowed to join. Thus they had never shared in the sacrifices. Instead of developing some commitment to the commune, they saw it as a means to their personal aims.

The thirty-year-old Duss set out to take control of the Society. According to him, Henrici and others had mismanaged Harmony. The commune allegedly owed more than the value of its assets.[58] Duss quickly got himself named as an heir with Henrici, who soon died.

Unlike most remaining members, Duss was fluent in English, and he easily diverted money to spend on his musical career. He toured the country with his band and even made appearances at the Metropolitan Opera House in New York City. Harmony had millions of dollars from sales of its scattered parcels of land in western Pennsylvania, which Duss seized to live a life of luxury.

When Duss withdrew from the Society in 1904, he received half a million dollars. Duss' wife, Susie, took control and continued selling Society assets. Only she and Franz Gillman, a mentally incompetent man, remained, and they formally dissolved the Society on December 15, 1905. Each received an equal share of over half a million dollars, and Gillman then gave his share to Susie Duss.

Society members had said the state should get their assets if the Second Coming did not occur before they died. However, Pennsylvania was slow in acting and got only the two city blocks that today comprise Old Economy Village in Ambridge. Twenty miles to the north at Zelienople, there is also a Harmony Museum with artifacts from the original 1804 settlement.

Bethel and Zoar Communes

Germans played an important part in several other religious communes.[59-61] Many of Count Leon's ex-Harmonists followed a German preacher, William Keil, to a commune he founded in 1844 at Bethel, Missouri. The Harmonists, as experienced craftsmen, helped Bethel set up factories and shops, a whiskey distillery, and a tannery for buckskin gloves.

After 1856, Keil and most of the 700 Bethel members, including the former Harmonists, went west to set up Aurora commune near Portland, Oregon. It prospered through hard work, but after Keil's death, no effective leader emerged. The 400 remaining members divided its thousands of acres and many buildings and stores. It had started with $30,000 in assets and ended in 1880 with $3,000,000 to distribute.[37,p158]

The people of Harmony and Bethel, with their strong religious feelings, work ethic, frugality, and their desire to perfect their lives apart from the world, were not unusual among immigrants to America. Swabia and nearby Alsace and Switzerland were hotbeds of religious dissent during and after the Reformation. The Amish and

Mennonites were groups formed there based on similar Anabaptist beliefs.

Another commune that stressed the Swabian values of spirituality and conscientiousness was Zoar. The name came from the biblical refuge Lot found after he fled Sodom. Zoar lasted eighty years.[62]

Joseph Bäumeler (known as Bimeler in America) was a pipe maker from Ulm born in 1778. Though of humble origin, he became a scholar and teacher with a gift for speaking and inspiring others. Like Rapp, he also embraced the teachings of the mystic Böhme.

Bimeler led Separatists to America in 1817. At first, many of the 300 immigrants from Württemberg worked as domestics for wealthy Philadelphians. Some became indentured servants to pay for their ocean passage. Eventually the group bought 5,500 acres sixty miles south of Cleveland.

Following Harmony's example, Zoar's members believed sharing in a commune would help them survive the rigors of frontier life. Their self-government was more democratic than at Harmony, with elected trustees and an elected council of five as a check on the trustees.

Zoarites, like the Harmonists, valued hard work, even on Sundays and holidays, including Christmas. They used spare time industriously. The women always knitted, even while preparing meals.

Commune members became laborers to help build a nearby canal. They also supplied food to the many canal workers. Zoar's main industry was farming. They soon began selling apples and cider, then beef, harness leather, butter, and cheese. The Zoarites then found iron ore, coal, and limestone on their land and built two blast furnaces. They hired outside managers and workers to help them learn how to work the foundry and make iron stoves, plowshares, and machinery. By 1835, less than twenty years after settlement, Zoar became almost self-sufficient in everything from food and clothing to building materials, furnishings, and tools.

The canal helped Zoar get into the shipping business. Again, Zoar first hired outsiders to teach them how to run the four canal boats, then taking over the business. The boats enabled Zoar to ship its bulkier products: ceramics, roof tiles, stoves, and other iron products, as well as agricultural produce. By 1852, the commune was worth a million dollars.

Modifying the Harmony Model

Zoar abolished sex as in Harmony. This fostered greater commitment to the group as a whole, but it was also useful economically. Women could do a full share of work, and the expense of children was avoided. But after ten years, cholera killed a third of the colony, and marital relations resumed. It is also said that another reason was that Bimeler couldn't resist the attractions of a young woman assigned to wait on him.

Bimeler was more enlightened and easygoing than Rapp. He was encouraging rather than critical like many Swabians. His severest punishment was exclusion from meetings. Bimeler wanted open minds and did not encourage indoctrination of the young. Eventually families could raise their children instead of the entire commune jointly doing this. Bimeler seemingly did not worry that children would form commitments to their families which would conflict with commitments to the commune.

Bimeler was also apparently not aware of the dangers of various extrinsic rewards sapping the intrinsic rewards that were the foundation of the commune. Members were allowed contact with American neighbors, which provided a way to make money on the side. The boys caught and sold fish, built and rented boats on the river, and slipped off to do odd jobs for outsiders. The waitresses at the hotel got tips from outside guests which they kept for themselves. Families sold surplus chickens and eggs to outsiders. The trustees tried to regulate some of this individualistic commerce, such as setting the numbers of chickens allowed families, but enforcement was difficult. People took in washing and sewing for visitors and boarders, made lace and other items for sale, and raised extra vegetables for market. The money was used to buy items not available at Zoar. Individualism was rampant.

When Bimeler died in 1853, the Society was worth over five million dollars. Unfortunately, no one else could match his charismatic spiritual leadership. Members thought of personal goals rather than the good of the commune as a whole. Many young people left for school or work. The Society's population dropped from a high of 500 before the Civil War to 222 in 1898. Some members complained they worked hard while others did not, yet everyone got the same rewards. Intrinsic commitment to the commune as a whole faded, and extrinsic rewards seemed to be more important for many.

A successful tourist business finally spelled the end of the Society. When a wealthy visitor retired at the hotel to enjoy parties and other entertainments, he won many members to his lifestyle. He was not religious. The amusements were a distraction to commitment to the commune as a whole. When Zoar was dissolved in 1898, it was still worth $3.5 million. Each adult received about $2,000 in cash, while the houses, land, and businesses were also divided. Many of the structures built by the Zoarites remain, some private and others open to the public as the Zoar Village State Memorial.[63]

As research has shown, the long-lived communes like Harmony, Bethel, and Zoar emphasized sacrifices that fostered strong commitment to the organization. Other communes without such sacrifices floundered because individuals were usually more interested in extrinsic personal commitments.[37]

3: FARMERS

A Swabian deacon visiting a parishioner in the hospital saw a bottle of schnapps by his bedside. The deacon remarked, "I hope this is not your only consolation." The sick man replied: "Don't worry, Herr Deacon, I've got six more bottles in the cellar at home."

In Swabia, as in the rest of Germany, alcohol was always popular. Most people drank their own wine and hard cider, and they brought this love of alcohol along to the New World.

The farmers in this chapter lived in the middle of Nebraska, near today's town of Eustis. Because they were so committed to a life of farming, they left Swabia for a new country with a new language and new customs. Farming promised them continued autonomy, a sense of efficacy, and the relatedness of the family members and neighbors who shared similar values. Most of the farmers had strong intrinsic motivation to live as a farmer. They also had to make major sacrifices to maintain their chosen way of life, and the acceptance of these sacrifices heightened their intrinsic motivation.

The early history of this settlement was collected in a book for the American Bicentennial.[64] My interviews, conducted in 1983, provide added insights.

Indians, Cattlemen, and Farmers

Indians occupied the land before the first settlers came. Between 1841 and 1866, some 350,000 people traversed Nebraska Territory along the Platte River Valley on their way to California or Oregon. At first, no one stopped because they considered the area uninhabitable. From about the 100th Meridian westward, the land was known as the "Great American Desert." It was a treeless, grassy prairie thought unfit for cultivation because of the sparse rainfall. Eustis was right at the 100th Meridian.[65]

The earliest records for the Eustis area cite trouble in 1864 when eleven were killed in an Indian attack on a covered wagon train. The cavalry came, built forts, and over the next fifteen years eliminated the Plains Indians, forcing survivors onto reservations.

Cattlemen then moved in. It was free-range country. Stock could be turned loose to graze. Farmers followed and fenced fields to protect their crops from the cattle. The cattlemen did everything they could to run off the early settlers. From 1878 to 1880, they often stampeded cattle over the farmers' small plots. Several long-range gun battles took place, but eventually the farmers defeated the ranchers after elections were instituted.

The almost free land attracted the farmers. According to the 1862 federal Homestead Act, citizens or applicants for citizenship who were at least twenty-one could take up to 160 acres for a filing fee of $10. After five years of cultivation, the land would be theirs free and clear. According to one Eustis homesteader, these requirements were loosely enforced. He only had to sleep on the homestead once every six months and bring ten new acres under cultivation every year. Some homestead land was still available in the Eustis area as late as 1892.

Chain Migration

The first immigrants who had successfully gotten their own farmland wrote ecstatic letters back to relatives and friends in the Old Country. That started an avalanche of immigration. According to the 1870 U.S. Census, two-thirds of Nebraska's people were either born abroad or the children of immigrants. There were so many Germans in Nebraska in 1867 the state legislative published the minutes of its first hearing in German.

Most immigrants from Germany were in their late teens or early twenties. Often younger brothers and sisters would follow at intervals of a year or so after putting in their time working on the family farm. Some newlyweds and couples with very young children emigrated as well.

Members of the Schmeekle family were among the first Swabians to come to the Eustis area. John, the oldest boy born in 1848, left the village of Einöd at nineteen. After spending two years in Illinois, he moved to Seward in eastern Nebraska, where there were many other German farmers. John married Henrietta Vogel there, and they started a family, later moving to the Eustis area.

Schmeekle's siblings soon came, first stopping to work on farms further east, then moving to the Eustis area. Later, others came directly to Nebraska. This "chain migration," where one person emigrates, writes back to family or friends who also emigrate and write back to others, was typical of German immigration in the nineteenth and early twentieth centuries.

One "push" for emigration occurred after 1871, when the militaristic Prussians absorbed Württemberg into the new German Empire. An important "pull" for emigration was the publicity by the railroads in Nebraska. Energetic construction of the transcontinental railroads began after the Civil War. By 1867, the Union Pacific's tracks crossed the entire state. The federal government gave the railroads half of the land that stretched twenty miles deep on both sides of their tracks. The Burlington and Missouri started a second line across the state in 1869. When it was completed, the two companies owned 15 percent of Nebraska, including much of the valuable river-bottom land.

Both companies printed special newspapers and brochures touting both homestead and railroad land. The publications were distributed in the East and in Europe. They also sent speakers on tours and stationed agents in European seaports including Hamburg. Burlington offered land with no payments until the fifth year and then at low interest rates. In 1870, railroad land ranged from $6 to $8 an acre. Twenty years later comparable acres still sold for only $25.[66,p165]

Shock of Prairie Life

The settlers' first reactions to arriving in the Eustis area were usually negative because of the barren prairie as far as the eye could see. Since there was no timber, the first shelters were often a dugout, a hole dug into the side of a ravine. Layers of brush and prairie sod on top served as a roof.

Later, sod houses were built by plowing up the prairie grass in foot-wide, four-inch deep furrows and then slicing across these strips every three feet or so. To keep these bricks of matted grass roots from breaking apart, they were laid on skids and pulled to the building site. The farmer then piled them crosswise, making walls three feet thick. The typical house dimensions were sixteen by twenty feet, with a sod roof laid atop pole rafters and brush. A man and a team could make a sod house in about three weeks.

Unfortunately, roofs often leaked after heavy rains. Dirty brown water got into everything. Even on dry days, dirt could sift down from the roof. The settlers could also find various insects, rodents, and snakes in their roofs and walls, and it was very dark inside.

Water was hard to get. In the East, men would dig a well fifteen or twenty feet deep, but in Nebraska the water table was often 200 feet down. It was incredibly dangerous for a man to dig a shaft so deep by hand.[65,p103]

Farming was difficult in Nebraska, too. "Breaking plows" could be as long as twelve feet with the front end resting on little wheels. Six yoke of oxen had to pull it. Only two and a half to three acres of the matted grass roots could be broken in a day.[67,p203]

Immigrants usually planted a vegetable garden first, then corn to feed the animals. Settlers chopped into the overturned sod and stuck corn kernels into the cut, stomped it shut, then prayed for rain. The prairies of central Nebraska averaged only twenty inches a year, often not enough for a good corn harvest.

The state geologist of Nebraska published an optimistic study in 1880. He claimed that cultivation would absorb rain and help produce more rain by releasing moisture from the ground to the air. It was said that "Rain follows the Plough."[68,p211] Scientists now know that turning over the soil brings subsoil moisture up to help growing plant roots. However, much of that moisture is lost through evaporation and from the strong prairie winds.

The geologist's study covered an unusually moist twenty-year period. Starting around 1886, drier conditions prevailed. The 1890s were grim. Twenty days in July 1890 had high temperatures above 100 degrees. Farmers could see the scorching winds wilt and kill corn plants before their eyes. Streams dried up and many wells went dry.

Religious people prayed in their churches for rain. Rainmakers came. Senator Leland Stanford of California recalled that during the construction of the Central Pacific Railroad they did much blasting in areas of the country with little rain. It rained every day when they were blasting but not after the blasting stopped. Congress appropriated money for experiments with explosives, but they were not successful. The railroads sent more rainmakers, and entrepreneurs came using secret chemicals, all with no more success.

Huge prairie fires also bedeviled the settlers. In the dry fall of 1888, most settlers did not yet have enough animals to graze. The dry prairie grasses were tall as a man. When farmers heard the roar of prairie fires miles away, they quickly had to plow a firebreak through the grass surrounding their homes.

There were also huge swarms of grasshoppers that ate the grain before it could be harvested. One infestation was so severe the Nebraska legislature enlisted men to eradicate them. To add to the distress of the nineties, the economic Panic of 1893 dropped prices for corn. At eight to fifteen cents a bushel, it was hardly worth harvesting.[65,p335]

The settlers stubbornly carried on. To feed their animals, they harvested prairie grass and planted oats. Winter wheat, which needed less rain and could be sold for cash, soon supplemented, then supplanted, corn.

At first, farmers sowed the wheat by hand, gathered it with a scythe, and used a flail to separate it from the stalk—the same primitive methods used in Germany since the Middle Ages. When a farmer saved enough money, he would buy a mower to cut grain, then a lister or planting machine, then a binder to cut and tie up shocks of grain. Eventually farmers used huge threshing machines, first horse-powered and then steam-powered, to separate grain from the stalks in the field. Investing in machinery paid off handsomely.

Tough Women

Threshing meant hard work for the women, who had to feed the large number of men involved. One settler, Kate Uebele, cooked for as many as twelve threshers: "We fed them breakfast, dinner, and supper, with lunch thrown in for good measure."[64,p10] Lunch referred to the midmorning and midafternoon meals, comparable to the Swabian *Vesper* eaten in the fields. A typical midday dinner could include noodle soup, chicken, mashed potatoes with gravy, green or yellow beans in gravy, tomatoes, cucumbers, lettuce, bread, butter, pie, and coffee. The typical *Vesper* in the fields included sausage, bread, butter, jelly, *Lebkuchen* (gingerbread), and beer or lemonade.[64,p11] Women might still be working at 11 p.m., to take care of laborers from far away who slept in the living room on a rug atop a pile of straw.

Women never rested. The men might spend two or three weeks building a sod house, but the women would clean the dirt sifting down from the ceiling for the next dozen years. The men might plow up a garden, but the women did most of the planting, weeding, and harvesting, as well as preparing and preserving food—salting beans and corn and drying them on the roof, pickling cucumbers in brine, making sauerkraut out of cabbage. This was besides their daily chores—milking cows, churning butter, baking bread, cooking meals, and caring for the children. They changed mattresses of straw and corncobs several times a year and plucked feathers and down from ducks and geese to make comforters and pillows. Women heated rainwater so they could wash clothes with a washboard, and then they wrung them out by hand. They sewed, knitted, patched, and mended clothes. In the fall, the corn-shucking mittens needed patching every morning.

As soon as they could manage it, children started helping around the house or with the animals. As in Swabia, both girls and boys helped with fieldwork—haying, shocking grain, and herding cattle. Teenagers often worked on neighboring farms and ranches. Introducing children early to farming probably contributed to their becoming committed to the farming way of life. Here, as well in most of the rest of the Middle West, German families continued to farm while others often sold out to seek the better opportunities to make money further west or in the cities.

The stubborn dedication to the farming life must have come about from a process of commitment much as with Conrad Weiser and the nineteenth-century communes. Life as a farmer can be hard, given the strenuous work at certain times of year, the natural disasters such as scorching high winds, drought, hail, wildfires, and pestilence, not to speak of the periodic collapses in prices for the farmer's products. If the farmers did not give up, and few did, they accepted the negatives, which only energized their commitment to the farming life.

Some farmers did give up. In one issue of the *Independent Era*, published some miles away in North Platte, there were fifty-two notices of sheriff's auctions or private sales. In 1894 there was a terrible drought, and corn harvests in Nebraska were 14 percent of normal. [65,pp346,349]

Making Alcohol

Farmers made wine for family and guests from the wild grapes that grew along the Platte River. For socializing, the men also went to a local saloon like in Germany where every village had one or more.

Christ Grabenstein's Empire Saloon was one of the first in Eustis. He had a cherry wood bar in front of a large plate glass mirror. Grabenstein got beer by train and kept the barrels in an icehouse. He also served liquor, mostly whiskey, and delivered beer within a sixty-mile radius around Eustis.

Alcohol was a traditional part of these immigrants' lives. Like Germans all over America, they fought against puritanical Yankees and religious people who saw alcohol as the work of the devil. In 1881, prohibitionists got the Nebraska legislature to set very high licensing fees for saloons. Grabenstein and others were not persuaded to close.

In 1910, Grabenstein got a letter from Carrie Nation, the muscular six-foot-tall woman whose husband had died of drinking. She had gotten a call from God to go to prohibitionist Kansas where, with her famous hatchet, she waged a crusade, smashing saloons. She warned Grabenstein she would be coming to Eustis to chop up his establishment and pour out his alcohol, but she was arrested in Kansas before reaching Eustis.[64]

Swabian culture emphasized the spiritual life, and they especially celebrated religious holidays, particularly Christmas. They imported the German custom of the Christmas tree, except in Nebraska it might be a small chokeberry bush. They also brought the German custom of the Easter bunny. The settlers dyed Easter eggs with coffee grounds and onion skins and held egg hunts. Many went to church on Green (Maundy) Thursday, Good Friday, Easter Sunday, and Easter Monday.

A German Methodist Church was founded in 1884, with meetings held in private homes. That same year, a pastor began riding out from another town to perform services every two weeks. Later that year, a regular minister was appointed. St. John's Evangelical Lutheran Church was established in 1887.

A Visit to Twentieth Century Eustis

A century after its founding, Eustis looked like many other Great Plains towns, rather dusty, with utilitarian buildings, nothing fancy. But for a town not near an interstate highway, it remained prosperous with a relatively steady population rather than the decline typical of many out-of-the-way small towns.

The Swabian mentality and culture had also endured even though the language and overt signs of the culture died out. Because the population of Eustis was homogeneous, traces of Swabian values and attitudes persisted.

When I interviewed residents of Eustis in the 1980s, a few children of the original pioneers were still living. These second-generation German Americans were often born in sod houses around 1900 or shortly after.

Cora Gruber was born in a sod house in 1903 and raised on a small farm that had a little of everything: milk cows, cattle, hogs, chickens, corn, and some wheat. Her family moved to Eustis so her father could run a dray, using horses and a wagon to deliver freight from the train station to the stores. The move made it possible for her to graduate from the high school in Eustis, the first in her family to do so. Many boys, like her husband, only completed eight grades.

Gruber told how her father came to America at age fourteen: "One of the farmers from Eustis went to visit Germany and brought him to America. Father said the man treated him so nice in the village. He got candy and everything, but once on the boat, everything changed. My mother was eighteen years old when she came over, and she worked for an English family before she married my father."

Gruber's parents spoke Swabian at home, but "I learned the American along with it, because my older sisters and brother spoke the American, too." She remembered how Americans treated the Germans badly during World War I. "The Moores ran the telephone switchboard. They weren't German, and they would listen in and say we can't speak German over the phone. but my father told them that he was going to talk German as long as it wasn't illegal."

Before the war, the immigrants felt pride in Germany, then arguably the most important industrial power in the world. During the war, Germans in America were forced to renounce their heritage and Americanize. The British controlled cable traffic across the Atlantic and spread propaganda like Germans eating Belgian babies. The propaganda brought America into the war and created enormous hatred toward German Americans, even precipitating a lynching in an Illinois town.[5]

A man who was a boy during that time in Eustis recalled: "Up until the war, we had a *Deutsche Tag* [German Day] down along the creek. There'd be military maneuvers with the German flag. They'd drill for thirty minutes. Then there was a special German speaker. That all ended."

His father refused to take him or the other children into town for fear they might say something in German. "A store clerk once spoke German to a customer who couldn't speak English, and they painted the clerk's house yellow." He added: "The Walkers were on our party line. They would pick up the phone when they heard the ring for our calls. They would listen, and if they heard German spoken they'd pound dishpans to drown out our talking."

Some pressures were official. "Our pastor preached in German, and the sheriff came and locked him up at the county seat. The deacons got him out so he could give confirmation. He had to promise not to speak German, and he stumbled through the confirmation in English. There were a lot of people in church who couldn't understand English."

After the war, the bad feelings toward Germans were rooted in economics. People in the Platte River Valley complained the Germans were moving in on them from the south and the Danish from the north. The sons of the German immigrants were coming of age, taking over farms around Eustis, and encroaching on the Platte Valley.

Stubborn Germans

People with Irish or English backgrounds often preferred to sell their farms for a quick profit rather than to hang on for the long haul. Extrinsic rewards sometimes took precedence over the intrinsic motivations for farming.

One Eustis resident told me the Irish or English "didn't look after their sons like the Germans did." The German settlers promised their children financial help if they wanted to farm, and most ended up farmers.

Eustis residents also claimed the Germans had an advantage when buying land. "The seller knew they would make it [not default on the mortgage]; they were frugal and worked harder. And when Germans gave their word in a sale, they were trusted."

After depressed agricultural prices in the 1920s, the economics of farming improved for a time. But during the Dust Bowl years of the 1930s, winds blew away a vast area of dried-out topsoil.

There were other calamities. "During the dirty thirties, we had migrating [grass]hoppers. They were dark green and just three-quarters of an inch long, not as big as an ordinary hopper. They'd come up from the south in a cloud a mile high that would make everything dark as night. Occasionally they'd come down and eat. One time they came when we had a twenty-acre field of barley that was not quite ripe. They cut about a third of the stems so the heads were lying on the ground, and then they just took off again. It looked like the whole earth was taking off."

Retired Eustis banker Herman Koch, a second-generation German American, told me how the family tradition of farming lasted for generations. The children were like their parents. "The farmers in Eustis are very conservative with regard to money." Local farmers always repaid loans unlike people in other towns. "People here do more thinking and soul-searching. They are more cautious than others when it comes to borrowing money. They figure it out on paper. If it won't come out right, they won't borrow the money."

He also stressed how carefully the German farmers spent money:

"They'd have a beautiful barn but just house enough to get by with. They'd say the house didn't earn any money." They drove Fords, the cheapest cars, not Dodges, Buicks, or Oldsmobiles.

Churchgoers

An outsider, not a German, who had married into a German family in Eustis, claimed the town was "the drinkingest, swearingest, most church-going place I've ever seen."

Cora Gruber recalled: "It used to be very strict on Sundays. No card playing was permitted, no baking. Your mother wouldn't even let you handle a pair of scissors. Of course, you had to be sensible. Our minister used to say that if you were driving a horse home and he fell in a hole, you wouldn't let him stay there until Monday."

Women's work was like pulling the horse out of the mud—okay on Sunday: "Everybody had company. People would drop in after church, a houseful of people, often so many people that we'd have two settings for dinner. The men would sit around and visit, but the women visited while they cooked, did the dishes and other work."

Religious training for the children was important. Gruber remembered: "Children would have the first two weeks of summer off, but then they'd go to Bible school for six weeks, from nine to four o'clock every day. My parents really stressed religion. We said prayers before bedtimes and meals, and we had readings from our German Bible." Religious German was High German, so with Swabian and English, people knew three different languages.

"The biggest share of people here are church-going. They observe the Sabbath. Maybe during the harvest time, they'll go back to work after church, but mostly they try not to work on a Sunday."

Alcohol and More Alcohol

As for the outsider's "drinkingest" perception of Eustis, another of the second generation Grubers, Paul, recalled that his grandfather brought a fifteen-gallon wine keg all the way from Germany. It was fitted with a harness so two men, grandfather and his oldest boy, could carry it everywhere.

Gruber remembered that every fall his father and his three brothers loaded their Model T with enough wild grapes to fill three fifty-gallon barrels. Others would fill a horse-drawn wagon. "We put the grapes in a big stone jar and used a wooden tree-trunk stomper to crush them. Some people used rubber boots. People might have four big fifty-gallon oak barrels of wine in their cellars."

Gruber remembered a lot of local stills, "Whiskey is easy to make; all it takes is sugar and corn. The first alcohol you distilled was really strong, 180 proof. You could use it for liniment. You could also dilute it and then add burnt sugar for whiskey coloring. We also used wild fruit for distilling—little round plums, oblong blue plums, and purple plums, any and all." This became plum brandy.

"The grown-ups didn't want the kids to know, so they'd put the stills in a chicken coop or an old shed. A feedlot was a very good place because you couldn't smell the still there. The alcohol was stored in gallon jugs or in some cases ten-gallon kegs that were buried for aging."

Eustis had a reputation. People would come from other towns, many of them dry, to drink in the saloons lining the main street. The train would even stop at the station a little longer, maybe twenty minutes, so passengers could step across the street for drinks.

Gruber recalled: "The tradition was that when boys were old enough to take care of the farm work on Saturday, fathers would head for the saloon and put a foot on the bar rail. A few stayed till Sunday." This was good for the farmer, but also for the boys, since they got a taste of the independence and sense of efficacy offered by farming.

After the repeal of Prohibition, Eustis businessmen quickly started serving beer again and opened a store that sold package liquor. However, in the 1940s the Women's Christian Temperance Union (WCTU) mounted a vigorous campaign.

I was told the WCTU emphasized one family where "the man is in every picture of a Eustis saloon I've ever seen. He was a regular fixture, bracing up the bar. An alcoholic, he was. That's the reason the women in that family were against liquor. And the sons, once they started drinking, they couldn't stop. So that man's wife was in the WCTU, and they really moved against hard liquor." The WCTU members were mostly non-Germans.

Other women recalled how a newly widowed wife found a cache of years and years of bottles in a hidden cave in the cellar. She had not suspected because she knew he was a good man who always went to church.

Humor

As for the perception of Eustis being the "swearingest" town, the farmers liked Swabian coarse humor. They will tell, for example, of a blind man who had a seeing-eye dog that was careless in leading him across the street, taking him into the thick of honking cars and screeching brakes. When the blind man reached the other side, he gave the dog a treat. "That's how I can tell which end is his mouth and which end is his ass, which I'm gonna kick."

Much humor took the form of practical jokes. The Eustis blacksmith used to suck the insides out of eggs through a small hole. When a fellow came by carrying a bag of eggs to take home, the blacksmith's cronies would distract him long enough to replace a few with the hollow shells. "Just think what his wife will say to him when she pulls those out."

Some tricks had a moral. Paul Gruber remembered when one of the Schmeekle family used to take a combine, some trucks, six men, and a cook to harvest wheat on land he owned in Colorado. Most were German, and each took a gallon of liquor to contribute to the group pool. But a man named Hausler took along an extra gallon for himself. When the others were working on the combine, he hid it under a pile of wheat straw, but the cook saw him. When Hausler was working far off, the cook got another gallon jug, filled it with water, and then broke it. When Hausler came back, the cook told everybody, "I went out shooting rabbits to fix us a meal. I thought I shot a rabbit under the straw, but it was a jug I hit." Hausler didn't say a word, and the others enjoyed his liquor.

Swabian food traditions also continued in Eustis. Some people made bread-like lye pretzels, *Spaetzle* or dumpling-like noodles, and *Schnitzbrot* (dried-fruit bread). Eustis celebrated a *Wurst Tag* (Sausage Day) festival every year with entertainment including different teams of men racing to see who could move an outhouse the fastest. Eustis population swelled to five times its normal 400 for the festival.

Male Chauvinism

The German men were often domineering, insensitive, and abusive to the women. Once a woman left home to get married, her parents would not let her come back home. She had to put up with her husband. It was said of a first-generation man, "It got really bad how he treated his wife. Some of her brothers had to go and say they'd lay their hands to him if he didn't stop abusing her."

The women had to make do with very little for the household and for themselves personally. Men worried, as the Swabian proverb puts it, "What the farmer brings into the barn with the wagon, the housewife carries out in her little apron"—that wives could fritter away men's hard work with many little expenditures. Men might caution their wives who wanted to buy a new table: "Make sure there's enough to put on the table." Asked to list her wedding presents, a second-generation woman responded that the interviewer was sitting on it. She added that her honeymoon trip was from church to her husband's home.

Koch, the second-generation banker noted: "In the German families, the man was always the head of the household. Among the old-timers, the husband did the business, and the wife didn't know if the family was worth ten cents or ten thousand dollars. There have been some sad cases where all of a sudden a family went broke, without any warning. Today, ninety percent of the women do the farm bookkeeping for their husbands."

Eustis men did not let women vote on church matters until World War II. "When the husband passed away, the wife was welcome to keep coming to church and paying to support it, but not to vote. Taking communion in church, the men always came first. That's the way it used to be for a lot of things. Eventually, ladies got on the town council here, while other towns had them there a long time earlier." Nearby towns were settled mostly by English or Scandinavians who had a more egalitarian culture.

Glennis, a third-generation Gruber, reflected on the modern changes: "Now, if a man would say 'do this' or 'do that" to a woman, she'd do the opposite. Women's attitudes have really changed. Back then, a woman wouldn't have talked back. I know when my grandfather had a few drinks he would physically abuse my grandmother. Now, a woman can make a living by herself. Parents used to encourage a daughter to marry a particular man for financial advantages. She couldn't go back to the parents."

A study in Swabia showed more depression among natives of Swabia than those born elsewhere in Germany, perhaps because of the culture's high standards and the harsh criticism of shortcomings.[69] Glennis felt that "a lot of woman suffered from depression. Men, too, for that matter. But a woman was home alone with a family to take care of, and everything was asked of the woman. The men were not good to these women."

Changes

The change from the subsistence economy of the early settlers to modern farming for cash crops has meant that women have to do less fieldwork. But many remain lonely with workaholic husbands who have no interest in a shared social life. As one woman put it, not entirely in jest: "About the only entertainment we get are the bull shows [auctions]. We try not to buy at the first show, so we can go to more."

While women have greater freedom now, the men of Eustis seemingly do not. A couple of second-generation Grubers, ranchers, recalled their immigrant father, like many Swabians, was free to develop rapport with his animals. In Swabia, if a peasant couldn't do a careful job with animals, he could make only half a living. They kept cows in barns all year and brought feed to them. They got to know the animals as individuals, and intuitive skills were developed.

In today's Nebraska, however, ranchers increasingly have such large herds grazing that it is not possible to get close to individual animals. Ranchers can become trapped in work that is boring, more like work on a factory assembly line.

The Gruber brothers found an outlet for their creativity. They started breeding longhorns as well as more modern types of cattle. Otto Gruber of nearby Elwood recalled, "It used to be that most all cattle were the same. A rancher raised shorthorns or Herefords. You used to go for looks on cattle. You bred them so they had a certain look, like a white face with white down the belly and a mostly red body. Now ranchers will have cattle that have all different color patterns, because it's the body shape that is really important. You want a long cow with lots of steaks. Now, you might cross Herefords with Angus and get a Black Baldie cross, and then cross those with a Charlois or Simmental."

The Downside to Modernity

Cora Gruber noted how farms were getting bigger to keep their expensive new machinery working all the time. "It used to be that each child here did better than his parents. There was a time when a farmer could accumulate land and give his children a chance. Now, just to keep afloat, a lot of farmers have to rent another half-section [320 acres] along with the half-section they own."

Larger farms have spread across America in part because they are thought to be more efficient, but agricultural economists have found that bigger farms make about the same net profit per acre as average-sized ones. One explanation is that these farmers have to hire less motivated, less careful employees to do much of the additional work.[70] However, farmers needed more and more acres for their incomes to keep up with inflation. They were forced to work longer hours, rent added land, and borrow more from bankers.

The greater acreage has meant problems for farmers in Eustis. In the past, neighbors or relatives on ranches would help one another on big jobs. However, with the increased emphasis on crop farming, everybody plowed, planted, or harvested at the same time, so they needed hired help, which meant pressures and responsibilities of supervising workers.

Farming became less of a craft or art where the farmer could take pride in his intuitive skills, like judging the value of a cow just by looking at it. The farmers used to enjoy fussing over their work, making careful decisions, and developing unique skills. Farming became more like monotonous factory work.

One old-timer recalled: "When you'd see a sunflower or cocklebur, you used to stop your machine, even in the middle of the field, and pull it out by hand. But now farmers spray ahead of time for weeds, and that's it."

Cora Gruber: "I think people were happier when I was young. Now farmers have to work half the night. You don't have the family life that you did then. When farmers used horses, they'd come home at the end of the day when the horses were tired. The horses knew when to stop. Now a farmer will work till eight or nine o'clock at night, and then the next one will take over the machinery and work at night using lights. The farms just have to get bigger and bigger these days, and where it'll stop, I don't know."

Not all the youths in Cora Gruber's generation stayed on their family's farm. They often transitioned from adolescence to adulthood by hiring out on the farms or ranches of neighbors. This would give them money but also an additional taste of the farming life.

One young man was able to go to school beyond the usual eighth grade. He came from a large family where the policy was that one boy would become a minister. However when his religious school folded, he went on to the University of Nebraska to become an electrical engineer, but such jobs were scarce during the Depression. He finally graduated as a teacher, but, he wanted more "action" than teaching high school math. "I'm an outdoors man." So when his father retired early, he became a farmer.

One of his brothers decided he'd get out of the house, work on the railroad, and make some money. He, too, was eventually drawn to farming like almost all of his peers.

Another of the second generation recalled how he got a teaching job in South Dakota. "But teaching didn't agree with me. I got thinner and thinner. Dad quit farming early and had a farm machinery business. He had two farms down in the Platte River valley, and that's how he helped me and my brother get started."

Growing up with Swabian values of independence and conscientiousness can make boys see that farming is compatible with the way they want to live. The farming life can be personally meaningful. When there is autonomy to choose the farming life, the resulting intrinsic motivation can be powerful. Not surprisingly, research shows that farmers who also work at non-farming jobs are more satisfied with the farming part of their lives.[71]

Too Much Drinking

Given the stress in farming, it was no wonder that some men intensified their traditional drinking. Research suggests that alcohol stimulates the production of adrenaline to cope with stress.[72] Researchers have also documented more drinking in rural areas of the United States, perhaps because it helps people cope with the hard rural life.[73]

Jim, a young fourth-generation Eustis resident, told me "maybe the farmers' drinking is a way of being tough. There are people here who are alcoholics, and they can still do the farm work. Quite a few of the Germans will drink even in the morning, but they can do the farming anyway. There's one guy who has really built up his farm, and yet he's never sober while he's working. You might see him drunker than a skunk, and two hours later, he's working hard. I'm not talking about people who are drinking away a farm they inherited. These are people who drink and work."

Jim recalled: "My father was harvesting corn in mid-November. It was late, and we were working fast. Well, he lost the ends of two of his fingers, from the joints to the tips, in the combine. So we finished the work for him. That was Friday evening, and the next workday, Monday, Dad was out working again. That happened two years ago, and even now, when he bumps those finger stumps, he jumps up in the air. You can see it took a lot of toughness to go out there and work after he lost those finger ends. Drinking can help farmers to keep going and do the tough work of farming."

Another theory is that alcohol provides a sense of power, especially in coping with a lack of control.[74, 75] Lack of power is apparently one explanation for the endemic binge drinking that used to be common in Ireland. Young men were economically dependent on their fathers for a long time, consequently turning to drink to cope with their powerlessness.[76, 77]

Other researchers believe that men who feel insecure about their sex roles drink more.[78] Alcohol produces feelings of strength and potency. Perhaps changes in the relations between the sexes might help explain some of the drinking in Eustis. Farm wives now have more education and greater status, often keeping complicated financial records for farms that are incorporated. Traditionally, German men ran the family, and this larger role for women must have made some of the older generation of men feel less powerful.

Recent changes in farming, to larger farms and raising single crops, may also foster drinking. Research shows that men with boring jobs drink more.[79]

Of course, all these sources of stress can contribute to the drinking. Men sometimes medicate depression from stress by drinking.[80]

The second-generation Eustis men hid their alcohol stills to avoid encouraging young people to drink. Apparently, as farming changed and life became more stressful, younger and younger people have started drinking. According to Jim, "Parents will even buy their children kegs of beer for graduation. Parents seem to push their kids to drink. Drinking in Eustis is very popular. There are always kids at the pool hall and the saloon."

A non-German from a neighboring town, one of many people from miles around who used to descend on Eustis as a kid to help the Germans celebrate their holidays, commented: "What is bad today is the number of young people in Eustis who drink. I have first-hand knowledge about alcoholics. I know what it's about, the different stages of drinking. I think if you check on Eustis a hundred years from now, you probably won't find the town any more. Today's kids are not growing up responsible enough to run a farm, and even if they were, they'd probably be too drunk to do it." Could it be that this opinion is colored a little by the stereotype that so many non-Germans have about alcohol and Germans—that Germans love alcohol too much?

Many of the small towns in Nebraska and others states are losing population. Farming is a hard life, and the attractions of larger towns and cities can't be denied. However, most of the young people in Eustis seem to be intrinsically motivated to live the farming life. Some may get jobs off the farm, but farming often draws them back.

Children are eased into farm work at an early age, often creating intrinsic motivation for farming. One man, an immigrant after World War II, tells how his five-year-old likes to help. "He watches the gate when we take the calves off the cows. He'll stand there so the calves don't come back through the gate."

A teenager tells of his positive attitude to farm work. "There's a tendency to want to work harder. On the farm, you can see the benefits of work. My dad teaches us kids how he does his business. He told us how he thought about the market and what we'd do with the cows and calves so we could eventually buy a better tractor in a couple of years." The father was showing how farm life can contribute to feelings of efficacy.

A fourth generation Gruber is a junior in high school and is leaning towards becoming a farmer. He works for his father on weekends and after school. "I'll drive a truck or fill a dryer. Maybe I'll water alfalfa by putting the aluminum pipes out in the field. The water never runs exactly right, so I'll have to move the pipes around to flood all the alfalfa. What I like about farming is the variety of work." As for his future, "It seems like college is a good idea—to keep up with new technology, to get better productivity out of a farm."

Some fathers caution their sons about going into farming. "I told my son he really should do something else. There are lots easier ways to earn a living, without the hassle and mental pressure of surviving from one year to the next. Well, he went into farming for the independence it gives him."

Most of the young men in the Eustis area probably will emulate their elders: In church on Sunday, swearing while they work on Monday, and taking a drink now and then, too.

4: EINSTEIN

Einstein, on his way home from New York's Penn Station asked the train conductor, "Does Princeton stop at this train?"

Some of Albert Einstein's peculiarities, such as his humble style of living, can be understood as an avoidance of the extrinsic rewards that can interfere with his intrinsic motivation to think about physics. He preferred old clothing like a comfortable mended sweater or shabby dressing gown. When he later visited President Roosevelt in the White House, he arrived sockless. He explained: "When I was young I found out that the big toe always ends up by making a hole in the sock. So I stopped wearing socks."[81,p27]

When Einstein visited the Sorbonne, the German ambassador to France had him stay at the embassy. The ambassador later complained that Einstein arrived with only a single pair of shoes, and "my valet has to clean them several times a day." Einstein in turn complained that his one pair of shoes was constantly disappearing. He would tell the valet not to polish them if it was raining. "They are going to get dirty right away, but he doesn't understand me."[82,pp27-28]

Einstein's simplifying of his life was reflective of his Swabian heritage with its frugality.

Swabian Origins

Einstein embodied the conscientiousness and asceticism common in Swabia. He used plain water instead of shaving cream to shave. Once he stayed in a four-room suite at the New York Waldorf-Astoria, but lived in just one room and closed off the other three. He had lost a tuxedo collar button and wanted to make it easier to find anything else he might lose. Einstein felt simplifying life increased his freedom. He wore his hair long to avoid going to the barber. He always wore the same leather jacket so he would not have to worry about buying another one. He eliminated unimportant things.

Einstein kept his distinctive Swabian accent all his life even though most other Germans saw Swabians as backward and stupid. Like a typical Swabian, Einstein also had a strong streak of individualism and his own brand of religion. Like pietistic Swabians, he considered work paramount. A biographer suggested that other Swabians would have also recognized Einstein as one of their own "in his speculative brooding, in his often roguish and occasionally coarse humor, and in his pronounced, individualistic obstinacy."[83,p8]

His father's ancestors had lived since 1665 in the small town of Buchau in Upper Swabia. Its Jewish community went back at least to 1382.[84] Einstein's mother was also Swabian, coming from Jebenhausen, near Göppingen. Even Einstein's second wife, his double cousin Elsa, was Swabian, from Hechingen.

Hell-Raiser

When he was young, Einstein was nothing like the mellow grandfatherly sage of his American years. He was a "rebel."[85] He was brash and arrogant towards other men, especially authorities, asserting: "Unthinking respect of authority is the greatest enemy of truth." He was sharp and insulting to those who displayed what he considered stupidity.[86,p6]

Early on, Einstein was full of himself. As a child, he attacked tutors and teachers. He hated having his autonomy restricted. Others thought he lacked discipline. Not so. He had self-discipline for carrying out his own ideas and spurned discipline imposed by authorities.

Einstein was always frank and direct, a common Swabian trait. To be polite can be considered lying. Even brutal honesty is valued because it promotes trust. However, Einstein was not malicious; he often tempered his directness with earthy Swabian humor.

In his youth, Einstein termed himself a "valiant Swabian" after the mythical hero described by the romantic Swabian poet Ludwig Uhland. In the poem of that title, a Swabian knight falls behind other Crusaders because his horse is weak. Fifty Turks then attack. When the first Turk comes within reach, the knight, with a mighty blow of his sword, slices him in half, cleaving skull, body, and saddle. The other Turks flee.

Asked about his deed by King Barbarossa, the hero explained that such "Swabian action" was common among his countrymen. Uhland, and Einstein too, understood that the term "Swabian action" was often used in a belittling sense to characterize the silly, stupid behaviors of the title characters in the Grimm Brothers tale, "The Seven Swabians." Einstein, like many Swabians, could poke fun at himself.[87,p263]

However, the label "valiant Swabian" aptly describes Einstein's brave attitude in dealing with the seemingly insurmountable problems he confronted as a young man. For two years after receiving his diploma from the Swiss Institute of Technology, he could not find a permanent job. He had to depend on a monthly allowance from his mother's family plus the pittance paid from private tutoring. He and his girlfriend, Mileva Maric, loved each other dearly but suffered long periods of separation. In addition, Einstein's mother vehemently opposed their relationship because Maric was Serbian and four years older than her son. Even after Maric had his baby, Einstein still could not find steady work.

During all this time he was engrossed in thinking about new ways to look at physical forces. It was this intrinsic motivation that kept him going despite the lack of money and other extrinsic privations. He could truly write Maric that he was being a "valiant Swabian."[88]

A need for autonomy and independence was an important part of Einstein's personality as it often is for Swabians.[5] Germans called him an *Einspänner*, one who travels alone.[89,p58] He admitted to being a "typical loner."[90,p11]

Einstein hated authority so much that later in life he wryly reflected that "to punish me for my contempt of authority, fate made me an authority myself."[91,p24] He also did not like to interfere with others' independence by telling them what to do. When another scientist once came to him for advice, Einstein sent him away, insisting, "I wanted to have time free for thinking, and I had no wish to dictate other people's actions."[92,p221]

Einstein loved music. Surveys in modern Germany show Swabians, compared to other Germans, are more likely to play musical instruments, belong to a singing society, and are more interested in various types of music including classical music.[5]

Einstein's mother, Pauline, was an excellent pianist and had him learn the violin when he was six. He hated his instructors, however, and only liked playing once he was thirteen and was free to teach himself. His second wife later described how "music helps him when he is thinking about his theories. He goes to his study, comes back, strikes a few chords on the piano, jots something down, returns to his study."[93,p301]

Einstein played the violin into old age and sometimes also the piano. Reportedly, he got more agitated over disputes with accompanists than with other scientists.[83] Playing music together cemented a friendship with the Queen of Belgium.

Childhood

Einstein was born in the Swabian city of Ulm in 1879, two years after his parents left rural Buchau. A year later, an uncle encouraged Einstein's father to move to Munich, where their firm made dynamos and measuring instruments for the new electrical industry. Like so many other assimilated Jews, the Einstein family valued material success and the respect of Gentiles as well as Jews. His parents were freethinkers who left behind the Orthodoxy of the rural Jews.

As have so many mothers of eminent men, Einstein's mother pushed him. She had him navigate one of Munich's busiest streets by himself when he was four. Perhaps his success in meeting this and other early challenges gave him the drive that would later lead to his bravely confronting difficult problems, when he had to be a "valiant Swabian."

Two childhood events were significant in Einstein's scientific development. When he was about four or five, he saw a compass and was amazed by the hidden forces in the world. He recalled, "this experience made a deep and lasting impression upon me."[94,p9]

A high school geometry text was a second major influence. The certainty of the proofs pleased him because, like most conscientious Swabians, he valued orderliness. He devoted his life to trying to uncover the hidden order in life. Unfortunately, he would later spend three decades in fruitless fighting against the uncertainties and disorder of quantum mechanics.

Religion

Einstein was a typical Swabian in focusing on practical things, but he also thought deeply about the forces behind them. For many Swabians, a personal God is behind life's events, but Einstein believed God acted through the laws of the universe. He felt that "nature is a magnificent structure that we can comprehend only very imperfectly, and that must fill a thinking person with a feeling of humility. This is a genuinely religious feeling."[95,p39] Einstein was like the eminent Swabian pastor Friedrich Christoph Oetinger who believed that God permeated the natural world.

Einstein saw no conflict between science and religion. Conflict came only from religious doctrines, symbols, and moralism. Einstein insisted that "morality is of the highest importance—but for us, not for God."[95,p66]

Einstein's religion consisted of "an attitude of cosmic awe and wonder and a devout humility before the harmony of nature."[96,p149] He revered transcendental mysteries and considered himself a deeply religious man. His religion concerned things beyond reality and almost unknowable, yet ultimately reasonable and beautiful. He felt that determined researchers were the only modern people who were deeply religious.[97]

In his combination of worldly and otherworldly concerns, Einstein took after the Swabian Pietists of the enlightenment, who built clocks and even primitive computers to calculate the time of the Second Coming. They also experimented with crops and fertilizers to discover God's plan on earth.

Not Good at School

Einstein's grades at high school in Munich were not good. His independence created problems with the authoritarian teachers. One teacher apparently complained of Einstein, "you sit there in the back row and smile, and that violates the feeling of respect which a teacher needs from his class." He was so disrespectful and disruptive that he was eventually expelled.[86,p64]

When the family's business failed and his parents moved to Italy, Einstein left Germany with no regrets. He was happy to avoid the compulsory military service.

Two years later, Einstein applied to the Swiss Institute of Technology in Zurich, considered the MIT of Europe. He wanted to become an electrical engineer but failed the entrance exam and had to spend a year at a provincial Swiss high school. A schoolmate saw the "impudent" Einstein as a "laughing philosopher" whose "witty mockery pitilessly lashed any conceit or pose."[98,p32]

After he was eventually admitted to the Institute of Technology, Einstein again had serious problems with authorities. He would throw a lab teacher's instructions into the wastebasket and do the assigned experiment in his own way. The last couple of years he stopped going to class.

One of his professors apparently complained, "You are a smart boy, Einstein, a very smart boy, but you have one great fault: you do not let yourself be told anything." This fault would serve him well.[86,p63]

Because Einstein had alienated faculty members who could have helped him after graduation, he could not get an academic position anywhere. He finally had to use a friend's connections to get a lowly job at the Swiss patent office.

There he enjoyed figuring out how the submissions worked and which deserved patents. He also had much free mental energy for reading and thinking about the mysteries of the physical world. His work on patents may have helped him work through his intuitions about basic physical forces.

Miracle Year

Still at the patent office at age twenty-six, Einstein had his "miracle year" of 1905 during which he got three highly significant papers published in the leading physics journal. Einstein made his scientific discoveries working independently in the backwater of Switzerland, without access to the latest scientific literature.

Working outside academia helped Einstein innovate. The typical European university of the time often discouraged radical new thinking. A single professor tightly ruled each department. On his own, Einstein could select useful ideas from different schools of thought.

Einstein worked in the tradition of the Swabian *Tüftler* or tinkerer, one who puzzles over problems and tries possible solutions. Rather than relying on sudden inspiration, Einstein, like most great scientists, immersed himself in a problem and tried out different solutions. He used different intuitions about the physical world to consider his new theories.

Einstein was a theoretical physicist, but he also had a thorough grounding in empirical physical reality. As a youth, Einstein was exposed to what his uncle and his father described from their electrical businesses in Munich and Italy. They made real things. They were not simply in commerce like most Jews in Germany at this time. Their company secured several patents, including a new technique for winding dynamo armatures, new meters to measure amperes and watts, and an arc lamp incorporating a mechanism to keep the arc constant despite carbon rods burning up. And during Einstein's work at the patent office, he got his own patent for a sensitive instrument to measure electricity.

Einstein's most unusual tinkering, however, was with his thought experiments. He did not devise elaborate laboratory experiments. Instead, he thought up imaginary situations to test his intuitions about forces in the physical world. Already at age sixteen, he used thought experiments to reveal that there was a basic contradiction between the theories of Isaac Newton and James Clark Maxwell.

Paradigm Shift: Relativity

In his work on relativity, Einstein had a big advantage over other scientists because he had correctly formulated a strategy: how to simplify electricity and magnetism into broad, unifying principles. Others were working on more specific problems. Einstein was working on a paradigm shift, much like an earlier Swabian, Johannes Kepler.

Until the seventeenth century, religion explained things in space. Kepler showed that geometric and mathematical laws explained how planets moved, opening a new way of looking at the universe. In turn, Einstein's new paradigm of relativity would shatter Kepler's mechanical view of the universe in which celestial bodies moved like objects through air.

In Einstein's new paradigm, he showed time and space were not fixed, but dependent on the standpoint of the observer. People found these ideas very strange.

Yet for Einstein his new theory was closer to the Swabian ideal of an ordered universe. Einstein's theory replaced the increasingly complex equations of physicists with his much simpler equations that described the invariance of God's plan. For physicists, invariance means that two observers will get the same results despite differing situations. All his life, Einstein would search for profound simplicities in physical phenomena.

Einstein built on James Clark Maxwell's newly developed

equations showing that electromagnetism consisted of waves that traveled at the speed of light—186,000 miles a second. Einstein thought that Maxwell had found an important invariance in nature.

In one of his thought experiments, Einstein imagined being on a train traveling at the speed of light and shining a flashlight down the train's corridor in the direction of travel. Could the light travel faster than that train? He didn't think so. A recent experiment had shown the speed of light didn't vary whether it was moving along or across the earth's path through space. Alternatively, Einstein considered whether a flashlight on a train traveling at the speed of light would emit no light because the light waves would be standing still. But this also made no sense.

Speed of Light as a Constant

In Einstein's thought experiment with the train, he considered that light should look and act the same to a physicist standing on a train moving at the speed of light and to another physicist watching from trackside. How could that be possible? Only if Newton's mechanical laws no longer held, only if time and space changed according to the speed one traveled! Instead of Newton's notions of absolute time and space, Einstein's science went beyond what we see and feel as reality.

Einstein decided that a clock would run more slowly as the train's speed approached that of light. The slowing clock would let the light of the flashlight appear to the observer on the train to zip down the corridor. Second, he decided that distances would change as the train approached the speed of light. The train would become compressed. The light waves from the flashlight will then have less distance to travel to hit the end of the coach.

The changes in both time and space made the light travel at the speed of light for the physicist on the train. Yet from the trackside perspective, the light traveled at the same speed—because clocks and distances there were "normal." For Einstein, the speed of light is constant independent of the speed of the source or of the observer.

By keeping the speed of light invariant, Einstein changed space and time into elastic variables. In other words, there is no absolute distance or time. A moving bar can be shorter than one at rest, and a moving clock can run slower than a stationary one. By rejecting the assumed independence of space and time, Einstein's relativity actually simplified equations earlier developed by physicists.

Of course, the relationships in Einstein's 1905 "special relativity" only held if both observers were moving at a constant rate. Other complications, such as differences in the force of gravity, would later be incorporated in his "general relativity" theory.

Another 1905 "miracle" paper had Einstein's famous $E=mc^2$ where mass and energy are different forms of the same thing. To come up with this equation, Einstein used a thought experiment where an atom undergoes radioactive decay with a slight loss of mass but a large increase of energy. In other words, matter was really 'lumped up" energy. This insight eventually became the basis for the atomic bomb. Hard for people to grasp at the time, Einstein saw these equations simply as features of God's orderly universe.

Einstein's third paper in the same year introduced the idea of independent tiny packets of light, later called photons. This insight was the beginning of quantum theory, opening the subatomic world to scientific study.

Using Planck's study of oscillators radiating energy, Einstein pictured a beer keg that could only be tapped for certain quantities of beer—liters, half-liters, and quarter-liters, and nothing between. Similarly, he decided, light and other forms of radiation came only in fixed, indivisible quanta or amounts of energy. In 1909, he suggested that each photon was accompanied by a wave field, an idea later known as wave-particle duality. For his discovery of quantum physics, Einstein was later awarded the Nobel Prize.

Einstein's miracle year of 1905 very quickly convinced top physicists that he had a first-class mind. Einstein had made significant breakthroughs in three major fields.

Swabian Modesty Out of Place

Four years after he had revolutionized physics, Einstein was finally able to quit his job at the patent office and take up a position at the University of Zurich. He remarked, "So now I am an official of the guild of whores."[83,251] Einstein still had his Swabian rudeness. It was said that "Einstein can express a strong dislike, and can fly into a passion, becoming intolerant and even unjust."[99,p77]

In 1914, Einstein went to Berlin and took up the highest university position in Germany. Despite this status, Einstein often behaved very modestly. He found it hard to accept his new lofty position. He felt "people generally overestimate me. I realize, of course, the value of my contributions to science, but I don't consider myself superior or different from any other men."[100,p29] He was modest in term of social status, not scientific matters. He could still be insulting to people who mistakenly claimed to have superior scientific knowledge.

Einstein's social modesty was common among Swabians, fitting their religious society. Since the Reformation, Württemberg's regimes had tried to develop biblical modesty among citizens. Since everybody had failings in the eyes of God, pride and arrogance were not appropriate. The last king of Württemberg was renowned for walking the streets alone to talk to citizens. To be pompous and act superior was not in accord with Swabia's egalitarian norms. The strong spirituality in Swabia meant that people tended to see everyone as equal under God. Everyone had God within. Furthermore, the religious courts had punished the possession of luxury goods that might make some people appear superior to others.

In addition, social classes were less important in Swabia than in other parts of Germany. There were few landed aristocrats. There was also almost no proletariat. Industrialization was widespread in northern Germany but had only begun to reach Swabia. Einstein claimed that "the distinctions separating the social classes are false, in the last analysis they rest on force."[90,p257]

The Swabian practice of sharing inheritance equally among male heirs had created a society of independent peasants where nobody had a basis for bossing anyone else around. Einstein's close friend, Philipp Frank, saw that "he behaved in the same way to everybody. The tone with which he talked to the leading officials of the university was the same as that with which he spoke to his grocer or to the scrubwoman in the laboratory."[101,p76]

However, after leaving Swabia for greater Germany, Einstein was thrust into a different culture. In the north, only the eldest or youngest son usually inherited the land, which created a class of wealthy farmers and a lower class of the disinherited. The contrast was most severe in eastern Germany, where the Prussian landowners had dispossessed the peasants who then became serfs or laborers with few rights.

Many of the dispossessed were drawn to Berlin, but so were the newly rich who benefited from the explosion of commerce and industry during the late nineteenth century. In Berlin, upper-class people protected their status, and lower-class people envied them.

Accompanying these status differences in northern Germany was Prussian assertiveness. The Prussians were originally a Baltic tribe that absorbed German culture and eventually conquered and dominated nearby Poland and then most of the northern German states. The Franco-Prussian war of 1870 and the humiliating defeat of France confirmed Prussian greatness. Prussia eventually expanded to take over all the German states except for Switzerland and Austria, creating the German Reich or Empire.[102]

The egalitarian Swabians often told disparaging stories about the Prussians. In 1849, the Prussians gained the Hollenzollern family's territory in the middle of Upper Swabia. They replaced local officials with Prussians. In an oft-told story, a child once accidentally spilled some water from a sprinkling can onto the dress of a judge's wife. The Prussian judge demanded formal apologies. When a newspaper in Württemberg picked up the story, Swabians there started wearing specially made watering-can pins just to infuriate the Prussian judge.

Problems in High Society

Given these cultural differences, it was no surprise that his Berlin colleagues thought Einstein a bit odd. As head of the Kaiser Wilhelm Institute, professor at the University of Berlin, and member of the prestigious Prussian Academy of Science, Einstein was an important person. Yet he tried to avoid putting people to extra trouble. A hostess once mistakenly gave him a fork for his pudding, and he used the handle to eat the pudding instead of asking her to get a spoon.

Berlin's high society expected that Einstein would, much as a newly arrived ambassador, make formal visits to other members of the Prussian Academy of Science. One morning while out walking, Einstein stopped to visit an academy member, but he was not at home. The maid asked Einstein if he wanted to leave a message, but Einstein said he'd continue his walk and come back later. When he returned at 2 p.m., the maid told him the professor had taken his dinner and was napping. Einstein left and returned two hours later, eventually to see the professor.

Einstein's casual clothing also got him into trouble. He didn't dress like a professor but like a simple pious Swabian. Once, when he was earlier a professor at the German University in Prague, Einstein showed up at a hotel for a grand reception in his honor dressed in the clothes usually worn by workers. Fortunately, the hotel staff eventually realized he was not an electrician.

Swabian modesty and plain dress were in line with John Calvin's teachings adopted in Swabia. Einstein was not concerned about influencing people with fancy clothing. He also avoided jewelry. Even worse, he would wear shoes with holes in the soles or a moth-eaten sweater. It was what he thought and did, not how he looked, that was important. Besides, dressing simply was economical. He once said the paper wrapped around meat should not be better than the meat inside.

Einstein was also modest about money. When he came to America in 1933, the head of the Institute for Advanced Studies asked his salary requirements. Einstein carefully considered his living expenses and proposed $3,000. When the director rejected this, Einstein asked, "Could I live on less?"[100,p37] The director countered with $17,000 and had to work to convince him to accept $16,000. The house Einstein bought in Princeton was much more modest than those of other professors. Apparently, Einstein would not visit Florida because he thought it was too luxurious a place. Similarly, his main meal was often pasta, and he would have sandwiches for supper, typical Swabian meals.

Einstein was especially critical of the vanity of other scientists. Einstein criticized Galileo's failure to recognize the prior work of Kepler. Unlike Isaac Newton, who would not publish work if he had to credit another scientist, Einstein made a special point of citing the work of others. He even thanked a co-worker at the patent office for help on his relativity paper. An observer saw Einstein as a "twentieth-century Ecclesiastes," quoting him as saying "Vanity of vanities, all is vanity."[103,p49]

Einstein also had some typical Swabian flaws. He could be sharply critical if someone did not meet his high standards for scientific excellence. As a young physicist, he scolded harmless critics. Einstein also had the Swabian love of earthy humor, reportedly cracking dirty jokes to his Berlin physics classes. Once, while visiting a summer home near a foul-smelling swamp, he said he was not bothered. "Sometimes I get even."[100,p56]

Swabians often relate better to things than to people, and Einstein's most important failing was his inability to be intimate with another person, perhaps because of the unpredictability of human relationships.

Marriages

Einstein had a pattern of seeking out motherly women and then discarding them when they became smothering.[104] Mileva Maric, a Serbian classmate at the Swiss Institute of Technology, fit this pattern. She was older, and Einstein's second wife was also a few years older than he. During their early years together, Maric reportedly contributed to many of Einstein's ideas.[105] It was eventually revealed that the two had a daughter out of wedlock who was first hidden away and then apparently given up to foster parents.

Maric's Serbian Orthodox parents at first opposed their marriage. After a son, Hans Albert, was born, Maric's father finally presented his son-in-law the dowry of $25,000, a fortune in those days. Einstein refused the money, saying he had married Mileva for love.

Because Swabian husbands and wives are typically so independent, strife is common and many Swabian jokes concern marital conflicts. In one example, a dying man confesses to his wife: "Many, many years ago, and only once, I was unfaithful with another woman." His wife replies: "I won't stand in the way of your dying in peace. I forgive you. But watch out if you want to get well again!"

In 1952, Einstein wrote to a friend: "I'm doing just fine, considering that I have triumphantly survived Nazism and two wives."[106,p14] He characterized marriage as "the unsuccessful attempt to make something lasting out of an incident."[90,p272]

Einstein openly discussed his many extramarital affairs with his second wife, Elsa, who reluctantly accepted. Not surprisingly, Einstein felt "if you want to live a happy life, tie it to a goal, not to people or objects."[10,p232]

At the end of his life, Einstein told the widow of his best friend that her husband had been able to do something that had always escaped him: live in a harmonious intimate relationship with a woman. Neither of Einstein's marriages had been free from conflict. He regarded this as the greatest failure of his life.

This difficulty was perhaps due in part to an obsessional pursuit of certainty. As a child, he had been taken with the definite requirements of the Jewish religion and, as a youth, the certainties of geometric proofs. Later, he would go over complex mathematical derivations and proofs with which he was already familiar. This was recreation.

Perhaps Einstein had fled intimate personal relations to take refuge in a physics where there can be more certainty than in dealing with people. The possibility of controlling matters is also often absent in interpersonal relations. His great need for certainty probably contributed to the later rejection of probabilistic quantum mechanics, which he felt was defective at its core. He admitted his difficulties in dealing with people, saying that "I have never belonged to my country, my home, my friends, or even my immediate family, with my whole heart."[107,p9] He later asserted that he was indifferent to the feelings of others but never as indifferent as they deserved.[99]

Obsessional commitments such as Einstein's to certainty are not uncommon. They apparently arise from unconscious needs or pressures, making the resulting behavior rigid and hard to change.[12] Picasso's obsessional work life was similar. He told his mistress, Françoise Gilot, "Everybody has the same energy potential. The average person wastes his in a dozen little ways. I bring mine to bear in one thing only: my painting, and everything is sacrificed to it—you and everyone else, myself included."[108,p201]

Steve Jobs is a more recent example of obsession. He had enormous intrinsic motivation, and his many innovative products made Apple the most profitable company in the world, but he was hell to work with.[109]

At least Einstein avoided telling other scientists what to do. He would send away those seeking advice, telling them to break their own eggs.

Women

Einstein took up his position in Berlin in April 1914. By the end of June, Maric and their two boys left Berlin for Switzerland. She had suffered increasingly from depression and serious physical pain.

Einstein had fallen in love with his divorced double cousin, Elsa Lowenthal née Einstein, who was living nearby. In a December 1918 deposition for their divorce case, Einstein stated, "My wife, the plaintiff, has known since the summer of 1914 that I have been in intimate relations with my cousin. She has indicated to me her indignation about this."[110,p349]

After the divorce, Einstein was not sure he wanted to be married again. However, to protect her reputation, Elsa's family insisted. They married in 1919.

Elsa tastefully furnished a Berlin apartment, and the couple had many visitors. One such guest, friend Charlie Chaplin, noted that Elsa "frankly enjoyed being the wife of the great man and made no attempt to hide the fact."[93,p301] Einstein liked being taken care of at home, but his wife and her daughters from an earlier marriage understood they needed to keep out of his way so he could think.

In a popular Swabian saying, women are a distraction from work: "schaffa, spara, Häusle baue, und net nach de Mädle schaua" (work, save, build a little house, and don't be looking at the girls). However, Einstein did have an eye for the women.

A student once saw Einstein obsessively cleaning his pipe and asked if he enjoyed that more than smoking. Einstein replied that pipes were like women: much suffering was necessary for a little pleasure.[104]

Fortunately, Einstein's second wife tolerated his many indiscretions. Einstein hired a young Austrian woman as a secretary at the physics institute and, with Elsa's permission, saw her twice a week during much of the year. While the Einstein family stayed at their summer home, she visited once a week to go sailing with him. In addition, an attractive wealthy widow sometimes sent a chauffeured car to pick up Einstein for theater dates and overnight trysts.

In a letter to a female friend who sought advice about her unfaithful husband, Einstein suggested that "a forced faithfulness is a bitter fruit for all concerned."[90,p119] Years later in America, Einstein gave the mother of a friend who cheated on his wife some philosophical advice: "the upper half plans and thinks, while the lower half determines our fate."[100,p135] In Einstein's case, the lower half sometimes overcame his formidable brain's judgment: he once propositioned one of Elsa's grown daughters!

Einstein had a close working relationship with Helen Dukas, who was his secretary for twenty-eight years. Einstein's son, Hans Albert, thought that his father was having an affair with Dukas. Her bedroom in their Princeton home was just off the study where Einstein spent most of his evenings and nights. When he died, Dukas inherited more money than Einstein's sons and sister. Dukas was also the sole beneficiary of the royalties from Einstein's many publications. In any event, she devoted most of her life to him and organized personal papers after his death.

Solving the Mystery of Gravity

During World War I, Einstein made great progress in his work. He generalized his earlier special relativity theory, bringing gravity into the theory. Special relativity only applied to observers moving relative to one another at constant velocities. It worked in the absence of all other forces, including gravity.

Previously, scientists had thought of gravity as a force that worked at a distance. Newton theorized the sun and Earth attracted each other according to their masses and the distance between them. However, the mysterious force of gravity worked instantaneously. This conflicted with Einstein's basic assumption that nothing could travel faster than the speed of light. It takes eight minutes for light to travel the 93 million miles from the Sun to Earth, so Einstein believed gravity could not be an instant pull.

In one of Einstein's many thought experiments, he imagined a box freely floating in space beyond the pull of Earth's gravity. Suppose rockets accelerated the box upward. Someone inside the box would feel his weight pressing down on the floor, as in a suddenly rising elevator. Also, a ball previously floating free of gravity would fall to the floor. Maybe gravity worked like such acceleration in space.

Einstein knew that all falling bodies always accelerate at the same rate, absent friction. So, he figured, gravity wasn't a force in the bodies themselves but instead in the space-time between them. What would it mean to be accelerated through space and time? Clocks would run progressively slower and objects or space itself would progressively foreshorten as the observer is accelerated. Of course, the mathematics was difficult, but Einstein's basic idea of gravity came from the simple elevator thought experiment.

Gravity was not some invisible force, but a change in space-time. Einstein counted it the happiest day of his life when he replaced Newtonian calculations with the space-time of his general relativity theory. With this theory of gravity as a warping or distortion of space-time, Einstein dismissed the flat-space geometry of Euclid that had ruled for two millennia.

Einstein could now explain the puzzling unusual movement of the planet Mercury near the Sun. In a 1916 paper, he proposed that warped space-time of gravity near the Sun could deflect starlight on its way to Earth. Three years later, a special British expedition saw his predicted bending of starlight during a full eclipse of the sun.

As a result, Einstein suddenly became world famous. He had made an enormous intellectual advance in understanding the universe. Far from being the threatening German, the stereotype during World War I, he was embraced by a war-weary world as a peaceful genius.

Criticizing America

In 1921, Chaim Weizmann, a Zionist who later became president of Israel, enlisted Einstein to play a key role in an American fund-raising campaign for a Jewish university in Palestine. Before, Einstein would not have associated with Zionism and its goal of creating a Jewish nation. It would have been out of character because he thought nationalism was dangerous. In addition, most German Jews, especially the Swabian Jews, were assimilated and well integrated into society. They did not favor a special nation for Jews.

However, in Berlin Einstein had seen many Polish Jews persecuted because of their strange dress, physical appearance, and use of Yiddish. Even German Jews shunned Polish Jews. Einstein's egalitarianism and his feelings for the persecuted underdog made him sympathetic to Weizmann's campaign for a Jewish university.

Einstein usually took pains to explain that he preferred cultural Zionism over nationalist Zionism. He would later write Weizmann that cooperation with the Arabs in Palestine was important. Near the end of his life, Einstein was offered the presidency of Israel, but he turned it down. He told his stepdaughter that if he were president, he would have to tell the Israeli people things they would not like to hear.

Einstein's trip with Weizmann raised some money, but on his return to Europe he frankly reported his negative impressions of America. Like so many Swabians, he wanted to be truthful and objective even if the truth might be unpleasant. He was not aware that Americans dislike criticism. Thus, he reported much prejudice in America against everything German, and even the suppression of the German language—which was not what he'd seen in England. There, people once again wanted friendly relations with Germany. The *New York Times* complained about Einstein's criticisms and attacked him.

Einstein also pointed out that Jews in America were not accepted into society as they were in Germany. He added further insults, saying how American women "spend money in a most immeasurable, illimitable way and wrap themselves in a fog of extravagance."[92,p349] Americans saw his comments as personal affronts. They didn't know that Swabians found Berliners' extravagances just as bad.

Other criticism by Einstein echoed Alexis de Tocqueville's nineteenth-century observation that Americans thought they were free but really just followed what others did.[111] They were less free than Europeans but unaware of this. It did not help that a German was criticizing Americans after some of them had died at German hands.

Pacifism

All his life, Einstein criticized warmongering. Many Swabians were pacifists because of the common belief the spirit of God is in every person. Taking a life is killing a part of God. Einstein had a deep emotional pacifism, declaring, "This is the way I feel because I find murder repulsive."[112,p151]

Einstein had arrived in Berlin in 1914 just before the beginning of World War I. The war was triggered when a Serbian nationalist assassinated the Archduke of Austro-Hungary in Sarajevo. Russia sided with Serbia, Germany with Austro-Hungary, and France with Russia. The British responded when Germany invaded Belgium to get at France. The British also saw Germany's growing industrial might as a serious threat, while the French sought revenge for their humiliating defeat in the Franco-Prussian war of 1870 and the loss of Alsace-Lorraine.

Arguing against war, Einstein signed a reply to the October 1914 "Manifesto to the Civilized World," in which ninety-three German intellectuals claimed that Germany was simply defending its culture. Only four signed the pacifist counter-manifesto, and it was never published. The war was popular among Germans, who expected a quick and glorious victory like the earlier triumph in the Franco-Prussian war.

The counter-manifesto predicted that all involved nations would pay a high price for the war and that no one would win. In 1914, German victory seemed certain, but Einstein called for a peace settlement brokered by world government and international cooperation, which he saw as solutions to war.

Einstein continued to oppose Germany's war efforts, and in 1916 the government banned him from making any further public statements. He turned to his scientific work for the duration of the war.

Germany was defeated and the Kaiser overthrown for a new republic. Though that pleased Einstein, he did not join the leftists trying to take power. At the University of Berlin, he spoke to the student council. They had deposed the rector and locked up the staff. He told them not to replace the old class-based tyranny of the right with a new tyranny of the left.

Later, while visiting the battlefield devastation in France, he said, "All the students of Germany must be brought here...so they can see how ugly war really is. People often have a wrong idea...from books."[92,p357]

Against Nationalism

After the war, Einstein became an enthusiastic supporter of the new League of Nations. He saw it as an ideal international vehicle to keep the peace.

In 1922, Einstein was invited to become a member of the Committee of Intellectual Cooperation, an organization similar to the later UNESCO. At the time, a new French government was increasing pressure on Germany, and French members of the committee objected to Einstein's presence simply because he was German. Ironically, many Germans felt Einstein, a Swiss Jew, shouldn't represent them. Einstein resigned from the body to protest the on-going French invasion and occupation of Germany's main industrial area in the Ruhr, but a year later, with the value of his renown recognized, he was invited to rejoin.

In an address at the German Reichstag, Einstein argued for European unity, but his advocacy of cooperation among nations was not popular in Germany. Later, Einstein wrote Sigmund Freud asking how psychoanalysis might help restrain man's tendency to make war. Freud replied that violence could only be overcome by a "community of interests" and the transfer of power to a larger unity "which is held together by emotional ties between its members."[113,p205]

Growing German nationalism and increasing Nazi street violence eventually challenged Einstein's pacifism. When Hitler took power in 1933, Einstein was on his way back from visiting the California Institute of Technology. He stopped in Belgium, where he was asked to defend two conscientious objectors. He surprised many pacifists by saying circumstances had changed with the rearming of Germany.

In a 1934 letter to a rabbi, Einstein claimed that he was still an ardent pacifist. But he didn't support refusing to serve in the military when aggressive dictatorships threatened democratic countries. He felt that a strong military can be dangerous to a nation's democracy, but serving in the Belgian, French, or British army would help protect democracy—and it would be foolish to wait until Germany attacked.

Much later, in 1953, Einstein succinctly characterized his pacifism: "I am a dedicated but not an absolute pacifist; this means that I am opposed to the use of force under any circumstances except when confronted by an enemy who pursues the destruction of life as an end in itself."[90,p161]

Personal threats played a role in Einstein's decision not to return to Germany. A Nazi magazine ran his picture with the caption "Not Yet Hanged."[114,p422] Einstein also heard rumors of a $5,000 bounty on his head. The Nazis ransacked his summer home outside Berlin and destroyed books and papers. This was because Einstein was a leftist. The Nazis put leftists in concentration camps even before they targeted Jews.

Refuge in America

Einstein finally accepted a position at the Institute for Advanced Study in Princeton, New Jersey, a post he held for the rest of his life. When American physicists became aware that Germany was trying to build an atomic bomb, they got Einstein's signature on a 1939 letter warning President Roosevelt. Einstein later regretted this. He felt it paved the way for the eventual dropping of atomic bombs on Hiroshima and Nagasaki in Japan, even though he had no part in their development.

During World War II, the Federal Bureau of Investigation felt that Einstein's "radical background" prevented his being a loyal American citizen and he should not be employed on secret war work.[90,p305] Nothing ever convinced Einstein that Hiroshima was excusable either morally or practically. He was probably not privy to American government estimates of millions of Japanese and American casualties in an invasion of Japan, but such knowledge probably wouldn't have changed his feelings. After the defeat of Germany and Japan, Einstein gave interviews and speeches stressing the need to internationalize the bomb and establish a world government to keep the peace.

Einstein spoke out in other ways that did not endear him to the American government. He said of the investigations by the House Un-American Activities Committee and Senate Internal Security Subcommittee during the McCarthy era that they were "an incomparably greater danger to our society than those few Communists in the country ever could be."[90,187] In 1950, Einstein wrote in a letter, "I have never been a Communist. But if I were, I would not be ashamed of it."[104,p60] His FBI file eventually totaled 1,427 pages.[114]

When the United States announced it was going to build a hydrogen bomb, Einstein went on television. He warned that further development of nuclear weapons would increase the chance they would be used and poison all life on earth. Unfortunately, his apocalyptic predictions, not uncommon among Swabians, had little effect on optimistic Americans.

Of course, Einstein's difficulties with the United States government were nothing compared with what he would have faced in Germany. In 1933, his books as a Jewish author were burned. The Nazis saw Einstein's theory of relativity as distorting the concepts of time and space. German physics was being attacked by degenerate "Jewish physics." Forget about the ten German Jews who had won Nobel Prizes for scientific achievement between 1905 and 1931.[115,p278]

A Gentle Clown?

Einstein seemed to change after he came to America. His sharp edges were smoothed. He appeared a wise and gentle clown. Perhaps he took on an otherworldly character so he could have the peace and quiet needed for continued thinking about physics.

In addition, some people around Einstein saw him as an embarrassment and un-American because he supported socialism, pacifism, and justice for African Americans. These were all radical ideas in the 1940s and 1950s. Einstein had already in 1938 become a charter member of Local 552 of the American Federation of Teachers in Princeton.

In 1949 Einstein published an article titled "Why Socialism" in the inaugural issue of *Monthly Review,* a socialist magazine. Einstein felt that ordinary working people would have no protection under unfettered capitalism, even in a democracy. Those with money would control politicians.

Einstein's handlers wanted him seen as harmless and politically naïve. The director of the Institute for Advanced Study opened his mail and turned down invitations without even consulting Einstein. Both during his life and afterwards, Helen Dukas covered up many potentially embarrassing facts, including the birth of his illegitimate child with Mileva, his messy personal life, and many romantic affairs.

Above All, Stubborn

The distinguished English scientist, C. P. Snow, once gave much thought about how best to characterize his friend Einstein. Snow came up terms such as "obstinate," "counter-suggestible," "independent," "non-conformist," and "deliberately impersonal." Finally, he settled on the term "unbudgeable" as descriptive of both his intellectual and his personal life.[116,pp4,18]

Einstein himself admitted that "God gave me the stubbornness of a mule."[106,p18] On one occasion, he and a colleague needed a paper clip. They found one, but it was bent out of shape. Eventually they located a box of new paper clips, and Einstein took one and shaped it into a tool for straightening the bent one. He explained, "Once I am set on a goal, it becomes difficult to deflect me."[117,p31] This single-mindedness about meeting challenges characterized his life as a whole.

In his scientific work, Einstein freely admitted his single-mindedness. As he saw it, the job of a scientist is to determine the most important problem and then to pursue the solution without deviation. Scientific greatness was a question of character. This included a refusal to compromise or to accept incomplete answers.

The Dream of Subsuming Quantum Theory

Many have judged Einstein's scientific work in the last three decades of his life to be self-destructive stubbornness in rejecting probabilistic quantum mechanics and trying to replace it with an all-encompassing theory. Yet these critics ignore that Einstein contributed a great deal to quantum theory. Instead of being the father of relativity, he should be considered the father of quantum theory. Einstein once told fellow Nobel laureate Otto Stern that he had thought a hundred times more about quantum problems than he had about relativity theory.[88]

In 1905 Einstein had identified the need for a new physics on the subatomic or quantum level. He was the first to recognize the gigantic theoretical leap made by discoveries at that level, and he was a driving force behind early quantum theory. He went on to introduce many revolutionary ideas that eventually contributed to the new theory of quantum mechanics. In 1916, he showed that electrons moved in different orbits around the nucleus in terms of probabilities—the orbits couldn't be pinned down exactly. And he later theorized the existence of new particles that we now call bosons.

Others went on to elaborate quantum mechanics, specifying how the subatomic world operated. According to Werner Heisenberg, both the position and momentum of wavelike quantum particles cannot be specified simultaneously. Uncertainty and probabilities are fundamental at the quantum level. However, Einstein insisted, "I shall never believe that God plays dice with the world."[118,p208] Physicist Neils Bohr supposedly suggested that Einstein stop telling God what to think.

The statistics-based predictions and lack of a physical explanation for what was happening in quantum mechanics distressed Einstein. While he was the first to see the probabilities inherent in quantum mechanics, he wanted causal principles. In quantum mechanics, events just happen. There is no certainty about what causes what.

Einstein wanted a science with more than empirical observations and mathematics. He needed a physical picture of precisely what occurs in individual atomic processes. He once explained his failure to solve a problem by saying, "I can't put my theory into words. I can only formulate it mathematically, and that's suspicious."[89,p28]

Einstein argued that quantum mechanics was only a provisional step to a correct theory. He believed there were hidden variables. When found, they would overcome the weirdness of Heisenberg's and Bohr's probabilistic theories. Einstein tried to include quantum mechanics and gravity together in a broader unified field theory.

Single-Minded to the End

During much of his early theorizing, Einstein almost always found that when his thinking was in conflict with conventional wisdom, his views were right. His single-mindedness was rewarded many times. So he continued until his death to seek a new scientific revolution like his earlier upset of Newtonian physics for relativity. Einstein spent the second half of his life looking for the rest of the story, the eternal truths that would explain all physical reality, from electromagnetism at the quantum level to gravity at the macro level. Many physicists felt the world's best mind was wasted in fighting against the progress in quantum mechanics. Science and technology passed Einstein by and developed all of our modern electronic miracles such as computer chips and lasers.

In Einstein's work on a unified field theory, a colleague recalled that "the search was not so much a search as a groping in the gloom of a mathematical jungle inadequately lit by physical intuition."[98,p356] Perhaps if Einstein had returned to his earlier method of using profound intuition and thought experiments to visualize physical forces, he might have been more successful.

One of the most amazing features of quantum mechanics is "entanglement." Interacting particles can continue to communicate instantaneously with one another despite moving great distances apart. Einstein discovered this in 1935. However, this instantaneous "knowing" or "feeling" at a distance violates the speed limit of light that is part of Einstein's relativity theory. So Einstein distrusted what he called "spooky actions at a distance."[98,p348] However, the phenomenon has been experimentally verified.[119] Also in 1935, Einstein calculated that black holes in space could come in pairs connected by shortcuts, or wormholes, another of his discoveries still not fully understood. Physicists are still grappling with the problems that preoccupied him in the last decades of his life, including the dark energy that seems to explain the expansion of the universe.

Work

To the end, Einstein followed the common belief of the Swabian Pietists that work is of ultimate importance. He told his son that only work gave substance to life. This importance of one's calling also helps explain much of Einstein's frugality and simplification of daily life. He recognized that extrinsic commitments were not useful and claimed that "the less that I can get along with in daily life, such as automobiles and socks, the freer I am from these drudgeries. If I don't have my hair cut, then I do not feel the need for a barber."[100,p111]

Einstein was clearly a workaholic. He sometimes even took a notebook along on his little sailboat. When the breeze momentarily fell, he took out the notebook and worked until the sails filled again.

How much did Einstein's two major lifelong commitments, seeking scientific truth and getting the world to embrace pacifism, involve sacrifices on his part, sacrifices that may have paradoxically strengthened his resolve?, The sacrifices for scientific truth started early, with his having to become a lowly patent clerk because he had been so headstrong in thinking in his own way. His pacifism at the beginning of World War I also resulted in sacrifice. The government muzzled all political statements for the duration. The same pacifism also meant that he couldn't participate with American scientific colleagues during and after World War II. Perhaps the greatest sacrifice was his later being ostracized by mainstream scientists for nearly thirty years after rejecting probabilistic quantum mechanics.

However, Einstein almost never faltered in his commitments, despite their costs. He once asserted the road to true human greatness led through suffering.

Einstein died in his sleep at age seventy-six from an abdominal aortic aneurysm. Found next to his hospital bed were calculations on a writing pad he had requested the previous day. To the very end, he had a single-minded commitment to solving the challenges of physical reality.

5: JEWS IN NAZI GERMANY

A Swabian joke tells how a wife kept her husband busy with chores during Saturday. Finally, she took a dust cloth from his hands and thanked him: "You've worked so hard all day. You can go out this evening with your friends." His reply: "I'm not going. I decided I won't take another order from you."

Some German Jews were similarly stubborn, rejecting all efforts by the Nazis to make them leave Germany.[9] By the 1930s, they had become more integrated into society than Jews in America. No quotas kept Jews from admission to German universities as they did in America. This integration had developed over the prior hundred years.

Forced Christianizing

In 1828, authorities in Württemberg, the core of Swabia, began efforts to assimilate the Jewish religion.[120, 121] They patterned the Royal Israelite Superior Church Authority after the similar Lutheran organization, with its president a government official. At first, he was a Christian, since Jews were not permitted to hold major government offices. Educated Jews occupied the other five positions in the authority.

This organization set job requirements for rabbis, who became state employees like Lutheran ministers. The new Jewish authorities dismissed most former rabbis because they lacked university educations. They also required new rabbis to give sermons in German.[120, 122, 123]

Three-quarters of Jews in mostly rural Württemberg were Orthodox, and initially they hated the Jewish Church Authority. And why should rabbis have to pass state exams?

Starting in 1838, Württemberg issued rules for synagogues. Jews had to enter synagogues quietly and with decorum and stay in their seats. Traditional practices were forbidden: pressing lips against the Torah, knocking during the reading of the book of Esther during Purim, lashes on the Eve of Atonement, beating of willow branches during the Feast of Tabernacles, and distribution of food during Simchat Torah. Violators were subject to punishment. Jews had to follow the orderly behavior of Christians in church.

At state-mandated synagogue services, the cantor had to sing in German before and after the sermon. Organs were also introduced. By 1841 prayers and sermons had to be in German, not the traditional Hebrew. New Jewish textbooks were patterned after Christian catechisms, with questions and answers to prepare for "confirmation."

Judaism was presented as a system of ethics and morality, and Biblical references were emphasized, as in the Protestant churches. Traditional Jewish laws and customs were seen as backward. Judaism became similar to Christianity in most respects. Jews now also had to observe military and other duties to the Kingdom of Württemberg. It was no surprise when Jews eventually came to say, "Stuttgart is our Jerusalem."[123,p132]

This Christianizing occurred all over Germany and also among German Jews in America. It eventually evolved into Reform Judaism.[124]

While German authorities felt important Jewish customs were disorderly, most symbolized the communal nature of Judaism—the Jews as a people who had a special joint relationship to God going back to Moses. Traditionally, the synagogue was the place where Jews could meet fellow Jews in both a spiritual and physical way, singing, dancing and eating together in celebration of certain holidays.

Many Jews tried to avoid the new strictures. City Jews who especially hated the organs in their synagogues fled to their villages where they could use unaccompanied voices during the High Holy Days.

Swabian Jews differed from Jews elsewhere in Germany in being mostly rural. The Duchy of Württemberg and the Swabian imperial cities had expelled them at the end of the fifteenth century, and nearby rural principalities had then welcomed them. This separation gradually ended after Napoleon merged the small principalities and the duchy of Württemberg into a single kingdom.

Rural Swabian Jews usually got along with their mostly pietistic Christian neighbors. But hostility could still erupt: pogroms against Jews occurred during economic bad times in the 1840s, including the failed harvest of 1846, and during the unrest of the failed German revolution of 1848. On Easter Monday in Baisingen, teenagers threw stones at Jewish homes, breaking windows and injuring people. They barged into some homes, smashing furniture and draining cider barrels.

There were also major riots in adjacent Baden, where the eminent liberal, Friedrich Hecker, lamented, "when we cuff the Jews we feel freer and stand higher." The peasants didn't have the vote, so they lashed out at the Jews.[123,p181]

In the first half of the nineteenth century in Württemberg, Jews in the small villages became wealthier than Gentiles. They also had more children, with the wealthier ones often leaving for the better opportunities in towns and cities. Children of the poor often went to America. The in-betweens stayed put in the rural villages.[123]

Bavarian Expulsions

Severe laws in Bavarian Swabia set quotas for Jews in villages: only one son could follow in his father's occupation and the others had to leave town; girls could marry only boys with the right of settlement and even then had to pay a high tax. As a result, many Jewish girls looked for marriage opportunities abroad.[125, 126]

The start of this emigration was noted in the *Israelitische Annalen* for 1840. In Ichenhausen, from a population of 200 Jewish families, sixty persons expected to leave. In Osterberg, twenty people wanted to go, out of a congregation of twenty-five families. The flight was so great the Jewish population in Swabian Bavaria dropped by one-third between 1847 and 1867.[125,p309] This exodus continued until there were only three thousand Jews left by 1910 and none during the Weimar times.[127, 128]

One of the emigrants was Bella Bloch, who was twenty when she left Osterberg in 1854 with her younger sister Fanny. Bella had been sent to Munich to learn the millinery trade so she could support herself in America. The Blochs settled in Newark, New Jersey, and three years later Bella's future husband, Gustav Kussy, came from Bohemia to Newark.[129]

In America, those in the Jewish community not related by blood or marriage often became acquainted at boarding houses where they took their meals and exchanged gossip, business ideas, and marriage possibilities. So it was for Bella and Gustav.

The Kussys opened a butcher shop and moved to the edge of Newark, where small houses were surrounded by gardens, farmland, and orchards. Settled by many German immigrants, the area became known as *Der deutsche Berg*, the German hill. German Jews often settled among German immigrants, since the common language was an advantage in business and socially.[124] By the late nineteenth century, Newark was about 15 percent German, with a third of the Germans from either Württemberg or Bavaria.[130,pp111-112]

Sarah, the Kussys' daughter, recalled picnics in parks and in beer gardens. Her parents would reminisce and discuss politics with German intellectuals, including the "forty-eighters" who had fled the failed liberal revolution. She recalled that her father bowled with neighbor Max Braun, the freethinking organist of St. Peter's Church, and with Father Prieth, the priest.

Most everyone was committed to speaking German. Synagogue services, as well as meetings of the woman's auxiliary at the synagogue and the religious school. Bible stories, the *Sidur* (prayer book), and *Khumesh* (Pentateuch) were all in German. Sarah Kussy recalled her Aunt Celia saying in German: "We are German, German is our mother-tongue, our children should know German."[129,p186] This emphasis on continuing the German language continued in America until the huge waves of Russian Jews arrived around 1900.[124,pp44-45] German Jews were very German.

With Germany's rapid industrialization after its unification, life for Jews got better and their emigration to America slowed to a trickle. In 1864, Württemberg issued a formal emancipation of the Jews, giving them the rights of other Germans. After 1870, Swabian Jews were citizens of the enlarged German Reich just like Gentiles. They left rural areas and took advantage of many new business opportunities, especially in northern Germany. The cities of Berlin, Hamburg, Frankfurt, Breslau, Leipzig, Cologne, and Munich had only 8 percent of all German Jews in 1850 but 50 percent by 1925.[131,p233] These cities, especially Berlin, also drew Jewish immigrants from Posen, the Polish part of Prussia.

In America and Germany, German Jews referred to the Posen Jews as "Pollacks" and looked down on them, since they were less assimilated to German culture. They spoke Yiddish and did not fit in so quickly. Actually, the average "Pollack" was, at least in America, more educated and had more familiarity with Jewish culture than most German Jews. Not being so "Christianized," they had superior religious training and in America they eventually became leaders in Jewish schools, charities, and other organizations.[132]

During the late nineteenth and early twentieth century, German Jews in America became outnumbered by the more than two million so-called "Russian Jews" from Russia, Russian-held Poland, Lithuania, and other parts of Eastern Europe. Having come from shtetls with dense Jewish populations, they were not as used to living among Gentiles, so many settled together in large cities, especially on the Lower East Side of New York City.

The arrival of Jews from Eastern Europe reduced the high rate of intermarriage between Jews and non-Jews. Between the years 1776 and 1840, nearly 30 percent of Jews in America married someone who was a non-Jew.[133,p25]

Life in the Villages

Swabian Jews were slower to leave the villages and small towns where they had lived since the end of the fifteenth century. In 1815, only 2 percent of Württemberg Jews lived in cities with more than 10,000 people. As late as 1925, 36 percent of the Jews in Württemberg were still in villages of less than 2,000.[125,p303] Swabia was densely covered by villages and small towns, and many Jews continued to choose a life integrated with the peasants, craftsmen, and small businessmen there.

Most of these Jews were cattle traders. Later they owned shops. However, many family members typically did at least some work in the fields, meadows, vegetable gardens, barns, or stables.[134, 135]

At the turn of the nineteenth century, the Jews and Gentiles in the villages "lived on terms of cordiality, if not of friendship."[136,p15] During the early twentieth century, every-day interactions between Jews and Gentiles in the villages increased. One man recalled that, "we all sang together whether rich or poor or Jew or Christian."[123,p248] The men also drank beer together in the evenings. Jews and Christians went to each other's weddings, and youths danced together on holidays of both religions. Jews and Gentiles belonged to the important veterans association. Retrospective accounts by Jews in villages of southwest Germany for the times before World War I were full of happy memories.[123, 135, 137] Like other villagers, Jews used the familiar form of "you" or *Du* for everyone in the village, Jewish or Gentile.

Jews were especially closely integrated into Christian life when they were young.[134, 138] The maid in Jewish homes was usually from a Christian peasant family. While Jewish fathers were away on business and mothers busy, the Jewish children often spent time at the maid's home where they would say grace and eat with the peasants. They would help with kitchen tasks, preparing the cattle feed, and milking the cows. They might help the family tie up hop vines, rake hay, pick plums, dig up potatoes, and thresh grain with a flail. On Sundays, the Jewish child was ceremoniously served the first piece of Swabian onion cake.

In Catholic areas, some Jews let the Catholic nuns educate their children. A Jewish child might be in the Christmas nativity play, and the parents would attend to see their child perform. Apart from worship practices and religious studies, the Jews and Christians mostly lived like each other. Later, when Jews moved to large towns and cities, they became secular Jews and only different from Gentiles by their observance of the high holy days. [123, 134]

Jewish Germans

By the time Nazis took power in 1933, most German Jews had conformed and become Germans like everyone else, socially, culturally, and even psychologically. Germans expected Jews to become German, and Jews mostly obliged.

Most Jews loved Germany. Several recalled for me in interviews that before the Nazis, "we were so German." They remembered how, "we were so assimilated." By 1927, 25 percent of Jewish men and 16 percent of Jewish women in Germany were marrying outside their religion.[139, 140,pp5,ii]

A teenager during the 1930s, Ella Heimann, told me: "We were dyed-in-the-wool Germans. My father was working in France, and he came back in 1914 to fight for Germany during World War I. He was the proudest German there ever was, and I was his daughter. I was German. I would have been happy to join the *Bund Deutscher Mädchen* [girls' Hitler Youth], if they would have let Jews in."

The *Bund Deutscher Mädchen* wore navy-blue skirts and brown Alpine jackets, the uniforms helping to foster a "we-feeling" among German girls—and a sense of not being German among Jewish girls.

Some Jews established a Jewish *Bund*. It was very German. The members recognized the "momentous achievement" of the Nazis and accepted the idea of a Third Reich that differentiated Jews strictly between those who were "German Jews" and those who were "Non-German Jews." The Jewish *Bund* only lasted a year or two before the realities of the Third Reich dispelled illusions.[141,p201]

Another teen, Gretl Temes, recalled for me the sudden changes after Hitler took power. She lived in the village of Buttenhausen and remembered mass demonstrations. "On the hills we would see fires where they gathered to sing hymns. It was mass hypnosis."

This was typical of the quasi-religious and apocalyptic mass movement developed by charismatic Hitler and the Nazis. There was a nationalistic liturgy, ceremonies, and rituals that appealed to the many Germans who were no longer religious. Hitler promised to deal with the catastrophes of defeat and territorial losses from World War I, the paralysis of the Weimar Republic's parliament, and the onset of the Great Depression. The Nazis would overcome the evil Jewish-Bolshevik-capitalist conspiracy and usher in a new world. The Jews were a convenient scapegoat, said to epitomize urban and modernist corruption, the opposite of Hitler's mythic view of the *Volk* (people) who were superior Aryans rooted in the ancient soil of Germany.[142, 143]

There were many different kinds of Jews in Germany, and the Nazis hated all. The intellectual, secular, and assimilated Jew was an example of detested modernity. The religious Orthodox Jew was the target of the traditional Christian hatred of those who killed Christ. The economically successful Jew epitomized money-grubbing capitalism. The socialist Jew represented hated Bolshevism and Marxism. Finally, the Jews who had arrived from Eastern Europe came with the alien culture of the ghettos. To the Nazis, these various identities made little difference. The Nazis hated an abstract "Jew" who lurked under all these different appearances. They were all members of an alien race.[144,p209]

After taking power in 1933, the Nazis quickly eliminated political enemies. Using the excuse of the Reichstag fire, Hitler had 100,000 mostly Communist and Socialist leaders arrested and sent to concentration camps. Other leftists either fled Germany or kept a low profile.

Most Jews, however, were not politically active, especially in the provincial backwater of Swabia, so they remained in Germany. Most simply saw Hitler as an incompetent. He would be just a temporary leader, like so many others during the Weimar Republic.

Government Officials Fired

The Nazis immediately began their work on the "Jewish Problem" by firing Jewish government officials. This was imposed with the "Law for the Restoration of the Professional Civil Service." Many Germans supported this. Some said, "Jews want to rule, not serve. Have you ever heard of a Jewish maid or a Jewish laundry woman?"[145,p29]

Many Germans accepted the Nazi propaganda that the Jews had been behind the "stab in the back" surrender at the end of World War I, the punitive Versailles Treaty, and the terrible inflation of the early 1920s. Germans had suffered, and it was now time for the people to come together and defeat their enemies, the Jews as well as the Marxists and the Allies who had prevented Germany's renewal.

Grete Marx told me the story of her husband, Karl Adler, a government employee. He was born in 1890 in Buttenhausen, just east of the Black Forest. Jews had lived there since the eighteenth century, and Adler's family for six generations.[146]

Like most Jews, Adler volunteered to fight during World War I, serving at the front, becoming an officer, and earning the Iron Cross. Grete told how Karl's father had also served in the army before World War I. "It was a matter of course to serve Germany, your homeland. Karl even volunteered to defend the king of Württemberg when Swabian Socialists threatened revolution after Germany's defeat in the war."[147]

Adler was the grandson of a cantor and trained as a singer and music teacher. He became a baritone soloist for the Württemberg court theater, but the war and severe head injuries cut short his career. After the war, a fellow officer had invited him to organize the music department of a new adult education program for Württemberg. With the advent of the Nazis in 1933, Adler was fired from his job as head of the Stuttgart music conservatory, a government job.

Eminent Jews like Adler also soon got special attention. After the elections of March 5, the Nazis became the largest party by far. The membership of the *Sturmabteilung*, or Brownshirt Storm Troopers, mushroomed, and this emboldened them to act against the Jews.

Adler got a phone call telling him a special delivery letter was on the way to him personally. When he answered the door, one man handed over the letter, while two others hit him over the head with iron pipes, knocking him unconscious. His wife recalled, "they didn't have uniforms and had caps pulled down over their faces, but we knew they were Nazis. All the men attacked that day were Jews." Adler went down bleeding, hit right next to his old battle wound. After his release from the hospital, the police offered to put him behind bars "for his own protection." Otherwise, they could not be responsible for his safety.

Instead of caving in, or running away, Adler became active in helping the increasingly oppressed Jewish community. As his brother-in-law, Walter Marx, recalled, "He was strong."

Adler developed a Jewish cultural organization to put on concerts, plays, lectures, and art exhibitions. Far from despairing at the increasing Nazi-ordered exclusions from public activities, many Jews found a new sense of pride and self-confidence in their own organizations. They felt Hitler could be thanked for rebuilding Jewish identity and solidarity. Zionism, belief in a needed Jewish homeland, started to become appealing.

However, Adler had to get all his music performances approved by the Gestapo (*Geheimstaatspolizei* or secret national police). They had started as a small Nazi police force that often worked outside the law. One day before an important concert, the Gestapo ordered Adler to remove a major work from the already-printed program. The text came from Psalms: "Oh, let not my foes destroy me!" Adler was not to say anything inflammatory when announcing the deletion, or he would be arrested directly from the podium.

Friends Change

Over time, Karl Adler and Grete Marx saw how their Gentile friends had to tone down their opposition to the Nazis. Pastors, as well as professors and schoolteachers, were all government employees and subject to orders from Berlin. They had to obey or lose their jobs. A close friend was the spouse of a Lutheran pastor who refused to take the oath that Hitler required of government employees. He resisted in other ways, and eventually committed suicide. Others opposed to the Nazis took notice.

Social relations with some Gentile friends of Adler and Marx continued. They might not shake hands or greet one another on the street, but many continued to see them at night. Neighbors would tell them the latest anti-Nazi jokes. A teacher at the conservatory, a former Communist, became a Hitler Youth leader, but he still showed the couple his latest musical compositions. They were stirring Communist songs to which he gave new, Nazi-approved wording. Many Communists switched to the Nazis after 1933 in the belief that anything was better than the chaotic Weimar regime.[148]

Much later, when the Nazi government outlawed Jews from buying certain scarce foods, the grocer would loudly warn Grete Marx in front of other customers: "You Jews, I can't give you coffee anymore." When she would get home, she would find coffee in her bag, with eggs and other items prohibited to Jews. Neighbors also brought them illegal foods.

These signals from sympathetic "Aryans" (pure, non-Jewish Germans) could give some Jews the feeling that things were not so bad. They couldn't believe that Germans would continue to accept the bizarre ideas of the Nazis about the Jews. Many had hope that things would change. They would wait and see. It couldn't get worse.

However, most Gentiles accepted the Nazi project to undo the "stab in the back" of World War I's defeat and the Versailles Treaty. People accepted that their duty as Germans was to follow Nazi ideas that the Jews had been at fault.

Grete Marx had befriended a young, sickly-looking man who attended the conservatory. She learned he came from a poor family in a small town, and she invited him to come for lunch once a week. Suddenly, though, he acted as if he did not know them anymore.

Her son, Fritz, also suffered like many other Jewish children. While playing nearby, other children warned: "This is our woods. You don't have any business here, in a German woods."

Before the Nazis took power, a music professor at Tübingen University had asked Adler to direct his composition with the university orchestra. It received good reviews. The professor wrote and complimented Adler on how quickly he taught it and how skillfully the orchestra performed and interpreted it.

After Hitler became chancellor, Adler presented a program by modern composers, including Hindemith. The same professor wrote a scathing review for the Nazi newspaper, criticizing "the Jew, Adler" and using a favorite Nazi catchphrase, *zersetzende Einfluss*, or immoral influence, to describe Adler and his "modern stuff."

Artistic Freedom Is Gone

Other cultural leaders, even non-Jewish, had to make major concessions to remain in Nazi Germany. Thomas Naegele told me of his artist father's experiences. Reinhold Naegele was born in 1884 into a long line of rural peasants and craftsmen, including a blacksmith who used to craft artistic ironwork signs for local inns. Naegele had been trained as a decorative painter but was otherwise self-educated. His Jewish wife, Alice Noerdlinger, a doctor, supported him in his artistic career.[147]

The increasing marriages between Jews and Gentiles in the early twentieth century had resulted in an estimated 70,000 children who were "half-Jews"—like the three boys of Reinhold Naegele and Alice

Noerdlinger—and nearly as many "quarter Jews."

At the beginning of World War I, Naegele had produced a painting that expressed his hatred of war. A ravenous goddess of death, a snake, types up a list of fallen soldiers while their mothers, wives, and girlfriends tear out their hair while surrounded by the blood of loved ones. An adding machine keeps tally. Since the war was popular among Germans in 1914, many people disliked the painting. After the war, Naegele continued his imaginative critical commentaries. One painting showed the ominous clash of mobs as part of the familiar beloved Stuttgart cityscape.

Naegele often used contrasts or fantasy to show conflicting everyday realities. He attacked the moral decay of the Weimar Republic. People inclined to reflect on the taken-for-granted norms of Swabian society loved his biting, critical works of art.[149, 150]

When the Nazis took power, they warned Naegele about being critical of the Fatherland. He had earlier juxtaposed a swastika and an ass's head in one of his works. Eventually he turned to tame landscapes and nature studies.

Naegele sold few paintings. His wife supported him and their three boys until the Nazis stopped Jews from working as physicians. After that, friends formed a subscription club of several thousand people who would contribute money and then eventually receive a painting. From 1935 until he left Germany in 1939, Naegele produced an enormous number of paintings. He felt humiliated by having to ask customers in advance what subject they wanted him to paint and in what style, rather than relying on his own inspirations.

Reinhold Naegele and Alice Noerdlinger were in a "mixed marriage" as defined by the 1935 Nuremberg Racial Laws. Some "Aryans" in such marriages divorced their spouses, but not many.

By a Nazi ruling in April 1939, Naegele and Noerdlinger were in a "privileged" mixed marriage, since the male was "Aryan." If the female was the "Aryan," the Nazis ruled it a "non-privileged" mixed marriage and subject to more restrictions. For example, privileged families could remain in their homes, while non-privileged families were evicted if they lived among "Aryans." They had to move to designated Jewish apartment houses.[140]

The Nazi Vice Tightens

The Jews who were neither eminent nor government employees had few problems with the Nazis at first. Seven percent of the Jews of Baden-Württemberg did leave in 1933, especially if they were leftists. However, only four percent left in 1934 and three percent in 1935. Most rolled with the Nazis' punches.

Jews did not let up their commitment to Germany and allow extrinsic concerns interfere: threats, deadlines, directives, or punishments. Being German citizens was still "worth it."

The first widespread action by the Nazis had not even been official—the April 1, 1933, boycott of Jewish shops, businesses, and professional services. This was planned for one of the best business days of the year, the Saturday before Easter. Nazi Storm Troopers took positions in front of stores and smeared anti-Jewish slogans on windows.

The boycott lacked public enthusiasm. German employees in the Jewish-owned stores crossed the boycott line to go to work. Other Germans chose precisely that day to visit a Jewish grocer. The stock market fell because many Jewish stores had ownership by foreign creditors or German banks. The Nazis canceled the Storm Troopers' boycott on its first day.

After this, most Nazi actions against the Jews followed the German preference for making everything strictly legal. The only major exception came years later with the Crystal Night pogrom in 1938.[140]

The first anti-Jewish law was the April 7th dismissal of Jews from civil service positions. On the 25th, another law set a quota that limited Jews to 1.5 percent in secondary schools. On the 10th of May, there was a public burning of "un-German" literature by Nazi student leaders in Berlin. This included any book written or published by a Jew, living or dead, and by leftist enemies of the Nazis. The nineteenth-century German-Jewish poet, Heinrich Heine, had predicted that those who begin by burning books would end by burning people. On the 29th of September, a Reich Chamber of Culture was created to give the Nazis control of all German culture.

Like other Germans, Jews respected legal authority and mostly complied with the new laws. They felt that Germany was a civilized country and the Nazi madness would pass. When Jews were later not allowed to use public beaches or baths, they usually obeyed. Jews evaded other laws, sometimes with the enforcers looking the other way, such as the prohibition against Jewish butchers doing kosher slaughtering. When Jewish youths could no longer attend secondary schools, their parents often sent them to Switzerland or other countries for further education.

After President Hindenburg died in 1934, Hitler combined the role of Nazi Party head with that of Chancellor, or head of state. From then on, he was the *Führer*, or supreme leader.

Early in 1935, the Nazis passed the Defense Law, making all "non-Aryans" ineligible to serve in the military. Patriotism was important to all Germans, so this clearly told Jews that they were not part of the Germany that they still loved.

The 1935 Nuremberg Racial Laws defined Jews as those with three or more Jewish grandparents, regardless of whether any had converted to Christianity. The law made illegal marriages and sexual relations between the races. To prevent so-called racial defilement, Jews could not hire non-Jewish females under age forty-five as household workers. Even innocent actions like inviting an Aryan girl to the movies could result in imprisonment, because this was an "erotic approach." It could lead to sexual relations.[115,p212]

The Nuremberg laws also underlined the second-class status of the Jews by stripping them of their German citizenship. The Nazis claimed they were simply restoring conditions existing before Jewish emancipation.

Despite the new restrictions, most Jews were satisfied enough with life in Germany to continue staying. America's quota for

German immigrants was still unfilled. Only another seven percent of Swabian Jews left in 1936 and the same percentage in 1937.

In 1938, the German government took a census of all Jewish property worth more than $2000. Compulsory Aryanization began, the forced sale of properties and businesses to non-Jews at depressed prices. Of the roughly 50,000 Jewish small businesses in Germany at the end of 1932, only 9,000 were still left by July 1938. Another law required Jewish males to adopt the extra first name of Israel, and women, Sarah. The purpose was to make Jews more visible.

Thirteen percent of the Jewish community left in that year.[145,p130] Finally, the extrinsic concerns of avoiding Nazi restrictions and abuse increasingly overwhelmed the intrinsic love of being a German citizen.

The Nazis started to force Jews out of jobs at Gentile organizations. Richard Heimann, a teenager in Stuttgart, told me how the Nazis wouldn't tolerate even lowly machinist apprentices like him. The Nazi national labor organization pressured his firm to fire all Jews. The company had a contract to train Heimann, so it told the Nazis that it could not let him go until the end of his training. The firm then stretched out the apprenticeship as long as possible, giving him two more months of work.

Many Gentile firms then refused to hire Heimann. He finally got a position with a Jewish firm. "I don't think they really needed me," he recalled. "I went there as a volunteer, and my father paid for my keep. The firm paid just pocket money and put me to work on something I wasn't really trained for."

Those best able to withstand Nazi pressures were self-employed businesspeople like Julius Marx, Grete Marx's brother, who told me his story. When the Nazis restricted the foreign currency needed to buy imported materials, Jewish firms turned to domestic substitutes. Some even went into different lines of business to avoid the restrictions.

After Julius Marx, born in 1893, had returned from military service in World War I, he married and moved with his wife, Liddy, to Neuffen to manage the family-owned factory. They were the only Jewish family in town but the most important employer, and they joined fully in the town's social life. Liddy did charity work, bringing poor people food and other necessities. She sometimes also took part in the haying or other fieldwork traditionally done by German peasant women. She and Julius also cultivated their own garden and orchard. They were very integrated into the rural community.[147]

Social Life Ends

Little by little, however, despite the Marx family's local prominence, formal social relations with Gentiles in Neuffen were ended. Nazi members of the singing society put on a Christmas play that was anti-Semitic, so Julius Marx resigned. In 1935, no organizations could mix Jewish and Gentile members. Segregation became law. The bowling club wrote Marx that he could still visit as a guest. He declined.

The Marx children were significantly affected, starting when all the other children joined the Hitler Youth. The schools also followed the Nazi line. As part of the 1935 May Day Festival, the students were to sing the Horst Wessel song venerating a Nazi martyr and "Jewish blood spurting from the knife." Only with great difficulty did Marx convince the staunch Nazi teacher to delete the song so his children could take part in the festival.

According to Marx, the *Landjäger* (provincial police officer) stationed in Neuffen was also "quite good, as long as he could be." One day, several Nazis noticed Marx leaving town with suitcases, and they became suspicious. They had the *Landjäger* contact the Ulm police to verify that Marx was simply going to nearby Herrlingen to attend his son's bar mitzvah at the Jewish school. The suitcases were gifts for his son, very useful to Jewish children during this time. To check, the Ulm police phoned Marx in Herrlingen.

Night of Violence

The bar mitzvah took place on the infamous *Kristallnacht*, or Crystal

Night as the Nazis called it for the glass shards from the destroyed windows of Jewish businesses.[151] Once Julius Marx learned that all Jewish men were being arrested, he quickly slipped out of the school's side entrance and made a visit to his mother in Cannstadt. There he called the *Landjäger* to say he would return to Neuffen the next morning. The response was: "That's all right, you don't have to report." Marx was one of the few Jewish men in Württemberg not arrested that night and sent to the concentration camp at Dachau.

Earlier, some Jews had welcomed legislation which Hitler termed a "once and for all" regulation of their position within Nazi Germany. This seemed to promise the end of the seemingly random laws and the brutalities.[152]

However, hopes were dashed with the nationwide Crystal Night pogrom of 9-10 November 1938. It was a turning point for Jews. The violence showed Jews that their self-identity as Germans was no longer valid. An important part of belonging to a nation, security and protection, no longer existed. The prior intrinsic motivation was gone.[153]

Nazis burned a thousand synagogues and vandalized Jewish businesses all over the country. Many Germans felt shame, shock at the extent of the action, and regret for the property destroyed. Many had still been shopping in Jewish department stores up to this point. However, Germans were afraid to intervene against the joint action by the police and Nazi Party. The German churches said not a word.[154]

A firefighter in Laupheim recalled that he was prevented from taking the fire engines out. Later, he was forbidden to pump any water until the synagogue had burned down. Meanwhile, marshals rounded up Jews and ordered them to kneel in front of the synagogue. "Then the arsonists came in their brown uniforms to admire the results of their destruction."[155]

This happened all over Germany. The American consul in Stuttgart reported the vast majority of the non-Jewish population evidenced disagreement with the actions and many hung their heads in shame. A small minority expressed satisfaction.[156]

In the concentration camps, the Jewish men were subjected to weeks or months of mistreatment and beatings. Some died. The sudden, unprecedented violence brought home to all Jews how serious the Nazis were about wanting them out of Germany. Wives asking about their imprisoned husbands were told they would be freed only if they could present immigration papers.

The Nazis decided the cost of the destruction would be charged to the Jews. They had to hand over almost 20 percent of the value of their assets, and insurance payments were also confiscated. Finally, all remaining Jewish businesses were to be Aryanized or liquidated.

The violence finally convinced most Jews to spare no efforts in getting out of Germany. In 1939, another 16 percent of Baden-Württemberg's Jewish community left. For the first time, the U.S. quota for immigrants from Germany was filled.

For most Jews, the gradual imposition of Nazi restrictions had been like the fabled frog being cooked on the stove. Eventually it would get so hot the frog might not be able to jump out of the pot. Fortunately, the abrupt violence of Crystal Night wakened most Jews—they should stop being stubborn, they had to get out of Germany.

The earlier Swabian tradition of emigration to America probably helped many Jews escape. A significant barrier was an American law, dating from 1882, that required immigrants get a sponsor to ensure they would not become public charges. With the onset of the Great Depression, this law was rigidly enforced. Fortunately, American Jews were ready to put their wealth on the line. Carl Laemmle, from Laupheim and a Hollywood film mogul, supplied affidavits of support for three hundred families.

Crystal Night ended Karl Adler's cultural programs. Three men came to arrest him at home, but he had rushed downtown to his office at the Jewish center, next door to the burning Stuttgart synagogue. The Gestapo found and placed him in a cell at their headquarters.

Early the next morning, a knock awakened Adler, his cell door opened, and a young woman came in with coffee and food. She told him: "Don't say anything. Just finish and give me the cup so I can get out without anyone seeing me." She was one of the secretaries who knew Adler from the times he had to report to the Gestapo about his activities.

Adler was then moved to a larger cell with other Jewish men also arrested on Crystal Night. The older men were especially shaken by their arrest and imprisonment. Adler helped revive their spirits by having each speak in turn. Then he led them in exercises.

Grete did not know where her husband was, so she went directly to the Gestapo. "Could I at least bring him some pajamas?" The official did not know where he was, saying in Swabian dialect, "Aber sischt ga nit nötig. Möglich er bracht's nimmer." (It is not necessary. Maybe he never needs them.) Grete did not know what that meant.

While other arrested men were transported to Dachau, the Gestapo released Adler after he agreed to three conditions. He had to stop all cultural work for the Jewish community, aid in the emigration of all Jews, and not try to leave Germany himself.

Collaboration with the Nazis

Adler was put in charge of a special governmental unit, the Jewish Relief Organization, with a staff of twenty. One department dealt with housing, since Jews were being moved out of mixed-race buildings and eventually out of many towns, to be concentrated in places that used to have Jewish ghettos. Adler also organized departments for employment, food rationing, and finance, the latter in charge of getting money from Jews to pay for the damage of Crystal Night and for other Nazi charges. Finally, a baggage department made sure that Jews took only permitted items out of Germany.

The Gestapo requested frequent reports from Adler on how many Jews had been helped to leave Germany. Unfortunately, at the same time Joseph Goebbels' Propaganda Ministry was proclaiming to the whole world that Jews were parasites and vermin. This made some countries reluctant to admit Jews. Even the Gestapo itself was not always consistent in actions furthering emigration. Contrary to the stereotype, Germany was not a well-oiled machine.[159]

Should Adler have refused to collaborate with the Nazis? As he saw it, he was suited for the job of helping Jews emigrate. Stuttgart did not have as many Nazis as other cities in Germany, and most of the pre-Nazi bureaucrats had remained on the job. They were local people who knew and respected Adler from his war service, his years working in a governmental capacity, and his work at the conservatory where their children had studied.

Adler felt he could take stronger stands because of his reputation. It was a dangerous job, for the Gestapo would hold him responsible for the actions of the entire Jewish community. But Adler knew he was the right person for the work, and someone had to do it. He felt it was his duty. Much criticism has been directed at later Jewish collaborators who facilitated shipment of Jews to death camps, not their emigration as was the case of Adler's collaboration.[160]

Grete Marx recalled how at first in Stuttgart it was a little better than in other large cities. "Then the Nazis put outside [non-Swabian] people in the Gestapo, SA [Storm Troopers], SD [Intelligence Agency], and SS [*Schutzstaffel*] They had to follow orders from Berlin, and they were not familiar with the local atmosphere."

The SS had originally been Hitler's elite guard. In 1934, during the so-called "Night of the Long Knives," Hitler had the leaders of the Storm Troopers murdered. He wanted his revolution to be legal, not based on the violence of the SA. He eliminated this alternative power center and set the SS in charge of policing the German state, with the Gestapo serving under the SS. The concentration camps and the later extermination camps were all headed by the SS.

The Gestapo

Despite their fearsome reputation, there were few members of the Gestapo. In 1937 the Gestapo used only 7,000 to police the sixty million people in Germany, and this number included secretaries and other assistants.

The Gestapo got information given by ordinary citizens who respected the laws and who would report illegal activity.[145, 161, 162] Germans wanted law and order enforced, and they believed bad deeds should be punished. Informing might also come for personal reasons—denouncing an enemy, a rival, or an envied person. Much information also came from the doctors, nurses, schoolteachers, and welfare officials who were all government employees and respected the law.

Whenever there was a report of suspicious activity, the Gestapo investigated. If, for example, a Gentile woman shook hands with a Jewish man, she would be interrogated about their relationship. Related information from other sources could lead to arrest without a warrant. Such actions created a climate of fear because of the many Nazi laws, regulations, and edicts. People became more careful. Some arrests were followed by torture and imprisonment at a concentration camp. Fear and the threat of violence plus a basic respect for authority meant that most Germans would police themselves. A massive police force was not necessary.

On one occasion, acting on tips from informants, the Gestapo asked Adler about the large amounts of money the Jewish Relief Organization was getting from an unknown source. Adler knew the source of the money was his longtime friend from the adult education movement, Hans Walz, a top executive with Bosch, a major Swabian firm.

On Crystal Night, Adler's wife, Grete Marx, had contacted Walz. The next morning a doctor brought her an envelope with several thousand marks to help get her husband and other men out of jail. Walz continued to send large amounts of money to finance the work of Jewish institutions such as nursing homes and orphanages and to help Jews emigrate. If Walz were found out, he would surely be put to death.

When the Gestapo chief asked where the Jews were getting

financial support, Adler replied: "You cannot ask me about that."

Then he was threatened: "Do you know what will happen, if you refuse?"

Adler cleverly responded: "Please listen. If I answer, I would be a traitor to my Jewish community who trust me. That is why I have this job. And I also have this job because you want to trust me." Adler looked him in the eyes and told him how his unique work as go-between would be threatened.

The Gestapo was known to shoot people in such situations, but Adler had chosen just the right words and attitude. The next time Adler was called in by the Gestapo, he was asked to seat himself. It was a sign that he had earned a little respect.

A Home Visit

Grete Marx also had a frightening encounter. Every year she made a *Succoth* bower for the traditional Jewish harvest festival. One day a Gestapo official who often dealt with Adler came to her door and asked if he could see the *Succoth* decorations. Grete wasn't properly dressed, so she had him and an accompanying junior officer wait downstairs. Fortunately, they did not notice the bookcases contained works forbidden by the Nazis.

They then went upstairs to the balcony, where the bower was built with branches from a nearby woods with decorations of fruits and vegetables. Grete answered their questions and told them that the absence of a roof symbolized how the Jews, and actually all people, are wanderers: "We have no permanence."

Grete had phoned Karl for help, and he arrived from his downtown office in ten minutes. He could explain more, since he had been raised as an Orthodox Jew and there had been a community *Succoth* bower in his hometown.

At the end, Adler added, "It is the custom that everyone who comes to *Succoth* receives something to eat." He offered the officers a piece of cake. The Gestapo men didn't sit down.

Adler then joked, "You can be sure there is no *Jugendblut*

[Christian child's blood] baked in it," referring to the common accusation against Jews that dated back to the Middle Ages.

The Gestapo officers eventually excused themselves. The visit may have just been a matter of simple curiosity and personal education, but you could not be sure.

Early in the summer of 1939, Karl and Grete sent their son Fritz to England with one of the many children's transports organized by the Jewish community. By this time, the Nazis said children leaving Germany could never return to visit their parents remaining in Germany. In August, Grete helped organize another children's transport and went along to visit Fritz. Karl came a few days later to get papers for their immigration.

Relatives in London tried to have the English police imprison Grete and Karl to prevent their return to Germany. But Karl pointed out that his parents were still in Stuttgart, and many of Grete's relatives were, too. Most importantly, Adler told his English relatives, "I gave my word to the Gestapo that I will come back, and I keep my word."

Finally Ready to Leave

The arrest and imprisonment of Jewish men on Crystal Night impelled many Jews to apply to emigrate in late 1938 and in 1939. Richard Heimann suffered terrible abuse for seven weeks at the concentration camp in Dachau. "We had five years to wake up to what was happening, but the arrest was, for me and many others, a salvation. The Nazis wanted people to realize that they meant business. They were serious about wanting Jews to leave Germany."

Heimann was lucky because his father had the foresight to file an immigration application for him at the U.S. Consulate in 1937. As his father told him, "It didn't cost anything, so I took out a number for you, too." It came up two years later.

Like many Germans, Heimann had distant relatives in America, descendants of his great-grandfather's brother. But they never responded to his pleas for help. Those who eventually sent affidavits of support were people who had left after World War I and then made a more recent visit to Germany. This personal contact made the difference.

Heimann understood why it was hard for people to give affidavits for someone they didn't know personally. "I would be very afraid. People can get sick or something. It could ruin you financially."

Unfortunately, too many Jews had stubbornly refused to cave in to Nazi pressures during the first years that Hitler was in power. In the five years since his taking power, only a little more than a quarter of the Jewish community in Swabia had left. With the violence of Crystal Night and the concentration camps, almost all Jews were finished with wanting to be Germans. However, American immigration quotas that had been unfilled for years were suddenly completely full. Most Jews wanted to leave but they could not.

The Holocaust

Time was now getting short for German Jews. In January 1939, two months after Crystal Night, Hitler had prophesied "the annihilation of the Jewish race in Europe" if "international finance Jewry" succeeded in "plunging the nations once more into a world war."[145,p141]

At first, the Nazis had ideas of resettling German Jews elsewhere, but Madagascar had to be ruled out because the British controlled the seas. The prime possibility seemed territory that would be conquered in Eastern Europe.

However, the Red Army put up strong resistance after Germany invaded the Soviet Union in June 1941. The first few months of the campaign cost the lives of over 100,000 German soldiers. The supposedly subhuman Russians fought fiercely. It seemed there would not be large new territories available to settle German Jews.

Another problem for the Nazis was the Atlantic Charter of August 1941 specifying an Anglo-American commitment to achieve "the final destruction of the Nazi tyranny." The Nazis thus expected the United States would enter the war one way or another. Hitler saw this as evidence of the international power of Jewish capital finally turning the world into an all-out struggle between Jews and Aryans. Allowing Jews to leave Germany would therefore be counterproductive. In addition, other countries were increasingly reluctant to admit Jewish refugees, contributing to the eventual German decision for a mass murder, a "Final Solution."[145,p203]

Apparently on September 17, 1941, Hitler decided not to wait until after a victory over the Soviets to deal with Jews unable to find places abroad. So on October 23, 1941, Gestapo chief Heinrich Müller passed on an order from Heinrich Himmler, his boss and head of the SS, that no more Jews could emigrate from Germany or from anywhere in occupied Europe.[163,p697] Instead, they would be deported to the East, to ghettos where Jewish residents had already been killed by mobile units of the SS and Security Police during the invasion of the Soviet Union.

By late 1941, almost two-thirds of the 10,000 Jews living in Baden-Württemberg in 1933 had been able to escape. More than half probably settled in America. But by 1945, of the remaining one third, only 616 survived the ensuing Holocaust.[157,p412, 158,p116]

Already in the few months before the end of 1941, 53,000 German and Austrian Jews were expelled from their homes and transported to the territories conquered in the East. In Stuttgart, the Nazis filmed the deportation of Jews to enshrine forever their great achievement.[145,p204]

People were told to prepare luggage useful for their "resettlement" in the East. Most believed they would begin a new life there, or at least they wanted to believe. However, even the prospect of resettlement could be hard for some to bear. As many as 10 percent of Württemberg Jews who received a deportation order in 1941 and 1942 were apparently so depressed they committed suicide.[145,p242]

The Nazis made the deportations and even the subsequent killing legal under various laws and regulations. These actions were carried out by subordinates who had received orders from duly constituted authorities. They were legal crimes. Most perpetrators were apparently not sadists or killers by nature. The Nazis made efforts to weed out those who took pleasure in the killing. Instead, the Nazis wanted people who could follow orders dispassionately.[164]

The view that one was obligated to follow legal orders to murder was not limited to Germany. In America, there was a major public outcry in 1971 after Lieutenant William Calley was convicted for the My Lai massacre of men, women, and children in a Vietnam village. Just after the conviction, 79 percent in a Gallup telephone poll disapproved of the verdict. People felt that Calley had just been following orders. Later detailed interviews found that 67 percent of Americans would follow orders in a similar situation. Many Americans felt that individuals were not personally responsible for actions when orders came from a superior.

Many Germans tried to remain ignorant of the slaughter of the Jews in the East from 1941 onward. They claimed they only knew that the Jews suddenly disappeared to go elsewhere. However, troops on leave in Germany told of witnessing the butchery of the Eastern European Jews. Most people in Germany had done nothing to prevent the earlier persecutions of the Jews, so admitting awareness of what the soldiers had revealed would probably have triggered feelings of responsibility and guilt.[154, 165,pp169,173]

Amazingly, once begun, the German determination to kill all Jews became obsessional. Eventually these efforts became so important that they took priority over German war efforts.[160]

Why Didn't They Leave Germany?

Why were some Jews eventually caught up in the Holocaust, ignoring the opportunities to go to America or elsewhere? First, during the eight years until 1941 the Nazis had tried mightily to force them to leave, Jews had no evidence that the Nazis were thinking of killing them. The Nazis had earlier carried out a euthanasia program for mentally ill and feebleminded Germans, but Jews were never seen in that way.

As mentioned earlier, Jews felt committed to Germany. The

country had been good to them, and they wanted to continue being part of it.

There were more specific reasons for not leaving. Some Jews, especially in rural Swabia, didn't think of leaving because they got along with their Gentile neighbors. Some of the peasants didn't accept Nazi anti-Semitic policies if they were applied too close to home. There were Swabian villages where Jews and Germans had lived together for generations.[154]

There were other reasons to avoid thinking about leaving Germany and starting a new life in a strange country. Some older Jews felt they were not a threat to the Nazis. They would be allowed to live out their lives peacefully.

Still other Jews felt they could continue resisting whatever the Nazis would do. They had survived so far, and they could continue coping. One man later recalled for me the German saying, "the soup when eaten is not as hot as when it was cooked."[115, 140, 154, 166]

Perhaps for some the strong commitment to Germany was obsessional. To think of leaving could be threatening. One would be giving up language, friends, and work—almost everything that was familiar. Perhaps the Jews continued to be firmly committed to Germany because the alternative was unthinkable. Maybe they loved Germany too much.[12]

Unfortunately for too long many Jews believed that the Nazis would be replaced by another, less hostile government. Jews hoped for, and saw, cracks in the Nazi regime. Starting as early as summer 1934, enthusiasm for the Nazis had started to wane for many Germans. Fewer people participated in rallies and meetings or watched parades. As German industry turned to financing re-armament, there were widespread complaints about shortages of food and consumer goods. Incomes after Nazi taxes and "contributions" didn't keep up with prices. Many people griped about the Nazis, if not Hitler. They thought things would change.

Eventually there were widespread rumors that the Catholic Church and German conservatives would get the military to depose the Nazis and even Hitler. From the beginning, the military had seen Hitler as a joke, terming him "the Bohemian Corporal."[148,p31]

Hitler was eventually forced to seek foreign triumphs to keep up his popularity and grip on the German people. In 1938 and 1939, he annexed Austria and the Czechoslovakian Sudetenland and then conquered what was left of Czechoslovakia.

Leaving in Time

Despite all the reasons for not leaving, many Jews did leave in time. Sometimes a wife secretly made arrangements to leave, and only then did her pig-headed husband give in to save the marriage. One couple traveled to Stuttgart for major dental work. As they were about to return home, the wife pulled out tickets for Switzerland and announced she was not going home again.

Women were more ready to leave because they were not as integrated into the public world. Men sometimes saw themselves as required to stay and resist. Men might also want to stay as long as they were still able to work and earn money. They didn't want to tear themselves away from their lifework, their clients, or their colleagues. Other men were too attached to the status and the possessions they had achieved. Finally, Jewish men felt more "German" and patriotic than women because of their greater education or earlier experience fighting for Germany.[140,p65]

Leaving Germany also meant giving up some important values. Gretl Temes told me how her father wouldn't leave even after spending six horrible weeks in Dachau. He wanted to follow proper legal procedures in selling his properties. Most Jews eventually sold for a pittance or walked away, but it took some time to accept that.

The son of a well-to-do businessman escaped to Czechoslovakia by telling German border guards that he was going only for a few hours. To convince them, he left his expensive car behind. He never came back.[5] After Crystal Night, however, border controls became stricter.

Jewish families scattered all over the world to whatever country would take them. Julius Marx's older brother, Leopold, sent his children to Palestine, and after Crystal Night, he and his wife went there, too. Julius Marx sent his oldest child to a boarding school in England and later sent a second child to America. The Nazis forced him to sell the Neuffen factory, and he moved the family to Cannstadt to the house where his mother still lived.

He experienced a typical bit of Nazi sadism during *Yom Kippur*, the Day of Atonement, the holiest day in the Jewish calendar. New regulations required Jews to turn in their radios on that very day. The September 1939 orders were meant to exclude Jews from the Aryan community which the Nazis had brought together through radio broadcasts.

Marx recalled how they took the heavy, bulky radios of that era to police headquarters in Stuttgart. "For transportation we had only a handcart. We were prohibited from using the streetcars. The pavement was rough cobblestones, so the radios were half destroyed when we got to police headquarters. The decree warned that if the radio was not working we would be charged for it. Then we had to wait for hours, still holding the heavy radio set in our hands. My very old uncle dropped his. In a way, he was happy."

The Nazis also built a concrete air raid bunker on the Marx property without even asking permission. Jews could not legally use these "German" bunkers during air raids. Fortunately, Julius and Liddy Marx and their remaining child were finally able to get American visas and leave in 1941 before there was any serious need for protection from British bombers. More important, they left before the Nazi decision in late 1941 that no more Jews could leave. Julius's elderly mother, however, did not leave in time. In 1942, she was shipped to the Theresienstadt concentration camp where she died.

Safety in England

Lilo Guggenheim, born in 1921, told me about tragic events just before the last escape routes for German Jews closed. Her wealthy family, part of the Jewish community in the industrial town of Göppingen, was reluctant to leave Germany. They had a good lifestyle, a mansion and a chauffeured car. When they finally accepted the need to leave, they contacted an elderly Englishman they had met on vacation. He agreed to accept teenaged Lilo as a domestic servant, and her parents hoped he would help them escape, too.

Still at home, eighteen-year-old Lilo wrote in her diary on July 12, 1939: "On the one hand, I want to get the permit [for England] fast, but on the other hand, I don't want to leave here at all, as long as my parents don't know where they will go. When I think 'War,' me in England and my parents here, I find it unthinkable." She left shortly afterward for England and persuaded her new employer to sponsor her parents.

Her mother wrote in reply to Lilo's letter home: "Now that Mr. Jacobs will become our sponsor, everything will end happily...Be a little flattering to Mr. Jacobs so he takes you into his heart." In late August, Lilo also got a letter from her aunt saying: "I am happy that you are with such good people, but who can resist you. Surely you do more than your duty...You are a magnificent girl, how you achieved to get a sponsor for your parents."

On September 1, 1941, Germany invaded Poland, and the British declared war on Germany. As an enemy alien, Lilo was prohibited from traveling more than five miles from home. She despaired and wished she were dead. She thought she would never marry and have children because they would suffer as Jews. "Today I am of the opinion that the Jewish Race should die."

In November, she felt even worse: "When I am alone with Mr. Jacobs, he wants to kiss me...He is a dirty old man...When I don't like someone, I turn cold...when I sat on his lap, he asked if my friend in Germany ever held me like that...I just remember what Mom said, 'flatter the old man a little, most old men like that.' Mommy, if only you knew how hard this is for me."

Matters changed quickly in Germany. Lilo's father, Julius

Guggenheim, was suspected of hiding gold in a shipping crate full of household goods. Lilo's mother, Lini, depressed and hopeless, suggested to her husband that they both commit suicide by jumping into the Neckar River. He made her promise not to do anything rash.

The contraband was found, and Julius was arrested. He heard Lini being questioned in the next room and hoped she would also be held overnight until their apartment would be searched. But Lini was released that same day, went home, and turned on the gas to commit suicide.

On Trial for Smuggling

Guggenheim's trial took place in January 1940. As the judges were assembling, the customs official came over to him in the dock and "squeezed my hand and encouraged me...I saw it as a friendly gesture and it definitely influenced the trial." Guggenheim defended himself: "I simply spoke in good Swabian [dialect], the way the words came out naturally." He argued that you cannot judge a deed by outward appearances without examining its motivation.

A child of poor parents, Guggenheim had worked his way up from apprentice to shopkeeper to owner of a large store. He interrupted his work to fight for Germany in World War I, earning the Iron Cross. He had to give up his store because of the war, and he lost his first wife during the 1918 influenza pandemic, leaving him alone with two small boys. He married Lini, a refugee from the former German province of Alsace. He resumed his business, but was wiped out during the hyperinflation of 1923. The third time he tried, he was successful, and by 1937 he employed two hundred people. He contributed to the welfare of his employees, not only financially, but by organizing excursions, giving them Christmas gifts, and so on.

After Crystal Night, he had been taken to a concentration camp and freed only because he had to help liquidate his businesses. He sent one son to South Africa, the other to America, and the daughter from his second marriage to England. He and his wife had moved to

Stuttgart and built a new house, hoping to retire there peacefully. He then tried to emigrate, getting visas for Cuba and selling his newly built home. But the visas turned out to be fraudulent, and the money was lost. "We had no more home as a result of the Jewish laws" and little income.

He told the judge: "I suddenly became afraid of the future. I was terrified...I would arrive in a strange country without help and without my own means...begging from strangers right away. In this desperate situation, I reached the decision to preserve some of my hard-earned money, which took me forty years to earn."

His argument appealed to the virtues of hard work, persistence, and self-reliance—all Swabian values. The judge acquitted him, and he was allowed to leave Germany.

Sad Leave-Taking

In July 1940, Karl Adler and Grete Marx were finally allowed to leave for America. Karl was a non-quota immigrant, because a friend had gotten him a teaching job at the New York College of Music. Their departure was very sad, for so many of their friends remained behind, unable to find a country to accept them. Letters asking Americans for help filled Karl and Grete's pockets.

As Karl's flight to Portugal was about to leave, the steward told him to get off the plane. He refused, because he could not leave his aged parents also on board to travel alone. When he eventually left the plane, he faced the commander of the airport who explained, "I saw your name on the passenger list, and I couldn't let you go without saying a word." He had been on the board of trustees of the conservatory, and his children had been students there, too. "I'm ashamed that all this happened, but I can't do anything. I just want to ask you not to forget the Germany we all have in our hearts." He then allowed Karl back on the plane.

Since they could not get a flight together, Grete had to take a separate flight to Portugal. "I thought I never will see Germany again, and I passed out on the plane."

Days later as their ship left, Karl and Grete watched the last lights of Lisbon disappear in the night. They thought of the words of their countryman, the Jewish poet Heinrich Heine, as he went into exile: "I once had a beautiful homeland."

"Then, of course, we didn't know what will happen in America, and if we ever, ever come back."

When Karl and Grete landed in America, they came with only $10 each. They were so embarrassed that they could not leave a tip on the ship. Germany had instituted a "Reich Flight Tax" to impoverish emigres. Starting in 1939, Jews could only buy dollars or other foreign currencies for 4 percent of their blocked German money.[140,p71]

At the dock in New York, they got a serenade from many of the people that Adler had helped flee Germany. They never saw son Fritz again after their visit to England. His ship to America was lost at sea without a trace. Grete's mother and uncle did not survive either. They hadn't believed they had to leave Germany, and they didn't get a number at the American consulate in time. Before Karl and Grete could send them Cuban visas, war broke out with Germany, and it was too late.

Escape from Germany was not the end of trials for many Jews, as Reinhold Naegele and Alice Noerdlinger learned. In the spring of 1939, Naegele finally realized that his wife was right about their needing to leave. They arranged to have his etchings and paintings, as well as their household goods, put into a shipping crate on September 10. Fortunately, a nephew living in East Prussia warned of the war preparations there. They left everything behind, including the paintings, and took the last train to Paris before Hitler invaded Poland on September 1.

The family thought they had found safety in England. However, after British forces were forced off the European continent at Dunkirk in May 1940, many Jewish refugees were interned as enemy aliens. The English soon released fifty-five-year-old Reinhold, but his seventeen-year-old son, Kaspar, was, like many other young Jewish refugees, imprisoned on the Isle of Man, later at a camp in Scotland,

and finally at another camp in Canada.[167] After a year and a half at prisoner-of-war camps controlled mostly by hardened and violent Nazi inmates, Kaspar was released. He eventually earned a doctorate at Harvard and helped establish the field of sociology in Canada.[168] He later tragically succumbed to depression and ended his life at age forty-two.[169]

Naegele's second son, Thomas, was younger so he avoided imprisonment in Britain and Canada, but he, too, found being a refugee painful. It took him a while to warm up to England, only to realize that what he did like there reminded him of Germany.

America: Pro and Con

Thomas Naegele reached America in September 1940. Others had told him America would be wonderful. Information would be freely available and dependable, not propaganda as in Germany. Institutions such as libraries and schools would not be segregated but open to everyone.

His first impressions of America were positive. Instead of the blackout he had experienced in England, lights blazed along Broadway. At Times Square, the Camel billboard puffed smoke rings nonstop. Automobiles seemed twice as big as those in Europe, and foods like hamburgers and orange juice were not rationed but sold everywhere.

But he was disgusted by Americans. "They seemed stupid about what was going on in Europe. They thought they could continue their lackadaisical existence. They didn't realize that President Roosevelt was sounding an alarm about the Nazis that could not be ignored. Americans seemed naïve. They had to have things explained to them."

Naegele also thought Americans lacked feeling. "I expected a more lofty generosity, and instead they were uptight about refugees. We were told we couldn't criticize anything, we had to register with the police wherever we went, and we couldn't travel out of a twenty-five-mile circle. The people all wanted to hear the same thing from us—what a wonderful place America was, how lucky we were to be here, and that we didn't dislike anything."

American attitudes changed when Hitler declared war on the United States in December 1941. Acceptance and assimilation into American life became easier for immigrants because workers were needed. Young men like Naegele were especially needed for the armed services.

His father continued painting after he arrived in America.[170] There was no market, however, for his symbolic critiques of society, especially from a German refugee. Nor was Naegele willing to make a radical change from the German expressionist style of painting that he had earlier practiced. He was stubborn. He did go to galleries and auctions, but when Americans started to turn to abstract expressionism after the war, Naegele was left behind. He never liked New York City. It was "hell."[171,p74]

Naegele gave his paintings to relatives and friends in exchange for help or hospitality. His wife again became the main support for the family. She did private nursing at first, then got her medical license and became a salaried physician at a transit camp for European refugee children going to adopting families. She eventually took a job as a doctor for schools in a poor section of the Bronx. Naegele again felt humiliated that his art was often paying off debts to others.

Returning to Germany

After the war, Reinhold's friend, the writer and publisher Josef Eberle, encouraged him to return to Germany.[172] Naegele visited in 1953 and found Stuttgart still in ruins from the bombing, but many people begged him to show his work again. American abstract expressionism had triumphed in much of the European art world, but Swabians were more down-to-earth and in tune with Naegele's commentaries on the social comedy of local life.

Naegele's 1954 Stuttgart show was an instant sellout. According to son Thomas, paintings he had trouble selling in America were almost ripped out of his hands. The President of Germany, the

Swabian Theodor Heuss, was among the thousands of visitors to view the exhibition. Naegele and Heuss had been friends since 1905.[149,p74]

Naegele found the interest and positive commentary on his work very stimulating. His paintings increased in value, especially since he was approaching eighty. Many thought it would be good to buy a Naegele before it was too late.

After his wife, Alice, died in 1961, he returned to live in his German hometown of Murrhardt. There he resumed working in an earlier style, reverse glass painting, and had a productive decade before his death in 1972.

Thomas Naegele served as a sergeant in the military police and guarded German prisoners of war. Back in America, he continued the family tradition in art, becoming a graphics designer for television and designing Christmas cards and U.S. postage stamps. He taught art in New York City but also made annual trips to Swabia, where he earned his own following for prints of rural life.

Some other older Jewish refugees also returned to live in Germany. After Julius Guggenheim remarried in America, he and his wife returned to live in the Stuttgart house he had built in 1937. Daughter Lilo spent the war years training as a nurse in London. She eventually became a dance teacher in America after the war, but she often vacationed in Germany at the resort where she had spent many happy times with her mother. [173] Karl Adler and Grete Marx did not move back to Germany, but they made several visits to close friends there.

Life for many refugees in America, especially older ones, was not easy. They had to live simply, as they had earlier lived simply in Swabia. Grete recalled, "We women dressed like our neighbors. We didn't stand out so much, like women in Nuremberg, Fürth, or Frankfurt. The Jewish women there dressed elegantly. I never wanted furs. That's Swabian maybe. You feel you don't need it in order to be something."

Karl Adler had been an important figure in Stuttgart, a major music center in Germany, particularly for Bach. In New York City, Adler became a small frog in a much bigger pond. He could not continue creating compositions that incorporated traditional Swabian folk tunes into classical works, since Americans didn't have the cultural background. Plus, choruses were not as important in America as in Germany. And Adler's Germanic style of teaching, sticking to one task and no fooling around, was also problematic in America. Some students called him a "German Nazi." He laughed at that.

Grete Marx recalled how they first visited Germany again in the 1950s, and she felt homesick. "I fell in love with it again, even though it was just a short visit."

Adler then received letters from some of the Gestapo men he had worked with in the Jewish Relief Organization. They asked Adler to vouch for their work in helping Jews emigrate. This might help them avoid conviction for war crimes.

Despite the blows to his career and the loss of his only child, Adler was still unwavering in doing right. He responded carefully, writing down exactly what they had done at the Jewish Relief Organization. They had gone on to Poland later, and he could not say anything about that.

Every year Grete and Karl used to get Christmas and New Year's cards as well as letters from one of the former Nazi officials. They wondered, was he an angel or a devil? Karl recalled that he was always correct. Perhaps that was why he had earlier not sat down or eaten anything at his visit to their *Succoth* bower. "Was it because there was another man with him who might reveal that he was sympathetic to the Jews?" Karl told Grete, "I imagine we could have talked reasonably together, and the next morning he could have ordered me shot."

Adler was unwavering in sticking to his values. On Crystal Night, the Nazis had searched his home and removed his ceremonial World War I officer's sword, considering it a weapon. It had a lot of symbolic value, and they later offered to return it, but he asked: "Do the other Jewish officers get their swords back, too? No? Then I don't want any exception made for me." He never got it back.

This strict adherence to principles was part of Adler's character, part of the religious and cultural heritage passed on for generations. Lilo's inability to respond to her employer's sexual overtures was also part of traditional Swabian values. Some would stereotype these Germans as rigid, but in such matters their inflexibility can be seen as a virtue.

To Israel?

Most of the refugees stubbornly hung on to their German values and culture. Most Swabian Jews went to America, not Israel. Zionism was never popular in Swabia or in Germany. Even Nazi oppression did not convince most German Jews that they needed their own state. They had been well integrated and committed to a modern state where everyone was treated equally under the law. Like other Germans, many Jews felt loyal to their country and to its traditions.

As Richard Heimann put it, "We certainly didn't want to be part of a Jewish nation that was to be created. Of course, being singled out and being resented made us feel more Jewish, but not more for a Jewish state." Heimann said he felt Jews were more a religious group than a people or a nation. His family had lived in the same village in Swabia for over 150 years, and he insisted: "I still don't feel a part of the Jewish people, of the Jewish state or nation."

In America, the Heimann family continued to be strongly religious, keeping up the traditions, ceremonies, and music they grew up with. Most of the synagogues they explored in New York City after first coming to America felt strange: Men and women sat separately. There was no music. The worshipers were less disciplined and orderly. "People start when they want to, and they end when they want to. They start their prayers when they come and then sing as if there was nobody else around. They talk with each other and then stand up and talk to God, all in their own time. German Jews are more disciplined, just like the Germans are. In Germany, people tried to follow the leader at the altar, instead of going at their separate speeds." The Heimann family found there were congregations such as Habonim in the so-called "Fourth Reich," the German-Jewish area of upper Manhattan, that continued the tradition of German liberal Judaism.[174]

Many German Jews thought ethics and morals were the most important part of the Jewish religion. According to their daughter Ruth Heimann, children learned the rules of behavior from their parents, which became habits. Eastern European Jews, by contrast, tended to feel that studying and debating the Torah were the most essential part of their religion.

Keeping the Old Customs

In America, the Heimann family maintained much from their former Swabian homeland: eating *Spätzle* (Swabian noodles), fermenting *Most* (homemade cider), staying put in the same house for a lifetime, even planting a linden tree out front. They had a painting by Naegele on the wall. They also visited Germany to see the friends they grew up with. These were people whose parents were not Nazis and with whom they felt they had much in common. "Our values are the important things, such as warmth, closeness, friendship, and peace in this world."

Many of the Jewish refugees continued to follow the values they learned from their elders in Swabia. Else Fuerst from the Black Forest recalled for me how her grandmother urged her to reflect at bedtime whether she had used her time wisely that day. This nightly review became a lifelong habit. She remembered, "You were not supposed to waste your time. We didn't play—when we had free time as young girls, we were sitting with *Handarbeit*, knitting or embroidery."

Another refugee, Ray Rothschild, told me that his parents always said, "*Gib mal Mühe*"—give it more effort, try harder. He became a millionaire in America, admitting, "I still go by that slogan."

Some continued to use the German language. Gretl Temes: "I taught my daughter only German here in America. I felt my English was lousy and she shouldn't learn wrong English. Of course, during World War II we didn't want to use German in the subways, so we didn't talk there."

Some Jewish Holocaust survivors can feel discomfort with German people, including even young present-day Germans or elderly Americans whose parents might have come from Germany in the 1920s. However, a study has shown that Holocaust survivors who were German-born were still positive towards German people.[175]

Acceptance of a German Heritage

Difficult as it may be to understand, even a survivor of the death camps could find it hard to reject his German upbringing and become vindictive toward Germans. In December 1941, and at age seventeen, Kurt Einstein was transported from Stuttgart with 1,000 Jews in cattle cars to Riga, Latvia. He was among the thirty who survived.

After the war, Einstein received a doctoral degree in behavioral sciences from Heidelberg. He underwent ten years of psychoanalysis and eventually became an expert in executive selection, heading his own large firm.[176]

Einstein told me that his mother had been "90 percent German and 10 percent Jewish. She was embarrassed to admit she was Jewish." This was common in Stuttgart.

In a speech in Munich to an organization of German and American CEOs, he described his terrible experiences. "I soon found out we were heading for death. Because I was young and able to work, I was shipped from one camp to another, seven altogether. In one of the camps, they selected ten of us every morning for hanging. We were forced to watch. Those who didn't watch went to the gallows, too. That devastating finger pointing at one after the other of the trembling lot of us still reappears in my dreams after more than forty years."

He added, "Looking at history, we Americans...had our own holocausts, we had slavery which in its basic form was no less cruel or devastating. The Romans, the Greeks...all had their holocausts, no less deadly or cruel than the Germans."

He concluded: "let us not bear a grudge or point a finger...Repressed hate has no redeeming character. We should point the finger at ourselves—every one of us could become an oppressor...forgive and never forget, and try to understand."

Einstein had no hatred towards the German people. He felt they were weak in succumbing to Hitler. He liked to go back to Germany and did so often. He also enjoyed speaking German, "even more so, Swabian." He enjoyed the Swabian mentality, how they can make fun of their own dumbness. He added, however, that they have a lack of sophistication and no understanding for complexities. He was somehow able to separate the Swabians from those who oppressed him. "My separation is on an emotional level, not rational." He even married a German woman, although the marriage did not last.

When he saw me out from his large Manhattan office suite, Einstein told a joke in Swabian. The many secretaries nearby would not have appreciated its extreme vulgarity, but they didn't understand Swabian either.

Ray Rothschild came to America in 1938 as a fifteen-year-old with his family and later was drafted into the army. Rothschild fought across Belgium and into Germany, where, as an officer, he was in charge of the occupation in a part of Hessia.

Before returning to America, Rothschild took an "Aryan" bride, even after learning that several of his close relatives had disappeared into the Nazi death camps. He felt comfortable with her, for both had in common commitments to German values and culture. Their marriage endured, while some of the marriages between non-Jewish American soldiers and German women did not.[177]

Many Jewish refugees returned to Germany on visits, sometimes after an official invitation from their former hometowns. While there, perhaps to pay respects at the grave of a parent, they found they still had emotional ties to former neighbors and colleagues.

Gretl Temes had left Swabia as a teenager, and she only returned on a visit with her sons because they insisted. They wanted to see where she grew up. When they went to her hometown of Buttenhausen, they intended only to pay respects at the Jewish cemetery. However, a former schoolmate recognized Temes and set up an impromptu reunion. She found it very gratifying.

In America, many of her Jewish friends, lacking a German heritage, couldn't understand how she could go back and enjoy herself with some Germans. Her ambivalent response was "we are forgiving, but not forgetting."

Her initial experiences with American Jews may also explain why she and other German Jews did not totally relinquish their German identity. In the 1930s, many American Jews hated all things German. Temes recalled, "We didn't speak Yiddish, so they were very suspicious." They felt, "What kind of Jew is this? You are a German!"

When asked why she met with Germans in her former hometown, she would tell others that they were kids. "They felt a lot of fear. That is why they did what they did. People were trained to inform on their families and on each other."

These refugees were all once German children, raised by their parents to have character and be true to themselves. Many German people have a romantic feeling towards their *Heimat*, or homeland, and some of the Jews driven out of Germany were not exceptions. It was not easy for them to give up their former commitment to being German.

The refugees went on to become committed Americans, too. They would say about America, "It couldn't happen here." They saw skinheads, American Nazis, and other hate groups as nutty people well under control by the authorities. The refugees were grateful to America, and they felt there was no reason why they should think otherwise. They were, in fact, good Americans who also preserved some of their German character. As another survivor of the death camps told me about his two American-born sons who had become rabbis, "Sometimes they are too stubborn. If the father is stubborn, why can't they be, too?"

6: PLAYWRIGHT BRECHT

Swabian saying: "Oh, if only people were as good as I could be!"[178]

Bertolt Brecht typified the conscientious and moralistic culture of Swabia. Like so many Swabians, he often criticized others for failing to meet high standards. In his life and his works, he often criticized society.

Much of Bertolt Brecht's eventual criticism from a leftist stance came from Swabian religious values. He came to see that the morality of the New Testament, particularly Jesus' concern for the oppressed, could be implemented with Communism—even though later in life, he was also critical of the Communists when they oppressed the common people.

Brecht did try mightily to get others to be good, and he never accepted that people often ignore New Testament virtues about treating others well. He was a consistent pacifist and therefore at odds with both Communist and capitalist warmongers. Unfortunately, his strict morality contributed to his failure in America. He did not accept that even in America, such biblical virtues can get little support.

A Softie Inside

Brecht was born in 1898 in Augsburg, the most important city in the Swabian part of Bavaria, fifty miles west of Munich. His mother came from a small village in Upper Swabia, part of a peasant family with deep roots in the area. In one of his poems, Brecht wrote of being carried in his mother's womb from the "Black Forests."[179,p3] His father's family was middle class and came from Achern, near the Black Forest part of Baden, with a culture similar to Swabia.

When he was young, Brecht would run home when other kids bothered him. He had heart trouble, was frail and nervous, and needed a night-light to sleep. He had a serious heart attack at age twelve.[180]

His mother, who also had health problems, spoiled him. She gave him license to go his own way and encouraged the growth of his self-confidence. The two became uncommonly close. She was well-read and felt he would be a great poet and consort with kings and princes.

Brecht adopted his mother's love of poetry, and she encouraged his writing, which would be intrinsically motivated the rest of his life. Brecht later recalled, "In poetry I began with songs accompanied by the guitar and composed the verses simultaneously with the music; the ballad was an ancient form."[181,p168] He would write more than 1,000 poems during his life, many of them providing important ingredients for his plays.[182]

Unlike many other Germans of the time, Bertolt didn't model himself on an authoritarian father.[183,pp18,120] Brecht was not close to his father, an important manager in the local paper mill. His father's limited attention was devoted to a younger son, Walter.

Fortunately, Brecht's father was mild mannered, and, according to a contemporary observer, "remarkably liberal in his attitudes." At worst, he would express amused disapproval or disappointment at Bertolt's rebelliousness. He once said about his son's poems and plays, "Personally I can't stand the stuff, but the youngsters seem to like it."[183,p18]

German culture in the early twentieth century was patriarchal.

Men strived to be strong and self-controlled. Brecht developed a hard masculine veneer to cover his anxieties and sensitivities. He was much taken by the saying "a warm heart beating in a cold person" as a possible recipe for greatness.[184,p185]

Brecht worried that his indignation over the world's injustices would wear off.[183] As a reminder to be enraged, he later hung on his wall a Japanese mask of a man with bulging veins at the temples.

Brecht also guarded his emotions by avoiding dependence on others. Like so many Swabians, and especially those with strong intrinsic motivation, he cherished his freedom and was impatient with any constraints, such as practicing his childhood music lessons. He showed his willfulness by always wanting to be in command, even when playing toy soldiers with others. His brother recalled, "His nature was always to boss others about, to impose his will on them."[180,p7] As a teenager, Brecht gathered admirers around him, so he could do as he pleased.

When Brecht dealt with authority, he used peasant-like cunning. Once, a teacher gave him a low grade on a paper. Brecht added more red marks and convinced the teacher that the paper's grade was wrongly marked too low.

Religious Origins of Brecht's Plays

Brecht was influenced by the Pietism still popular in Swabia during his youth. In 1928, after his *Threepenny Opera* became famous, Brecht admitted the Bible was the biggest inspiration for his writing.[179,p112] German schools included religion in the curriculum, and Brecht made much use of Luther's Bible.[185, 186] He got an extra dose of its language from his mother and grandmother who read Bible passages aloud to him. Many of the dialogues in Brecht's plays would later be biblical quotations and parodies.

Brecht accepted the fundamental values of Christianity, especially the problem of how to be good, but he rejected its transcendentalism and mysticism. He had an "unromantic realism."[183,p50] When he was later seriously ill in Berlin, a friend asked why his hospital bed showed that he was Catholic. Brecht explained the priest was friendlier than the Protestant chaplain. Brecht often turned religious teachings upside down. He rejected fatherly dogmatism, if the good motherly impulse was in conflict.

Brecht was an atheist. He apparently saw religion as a terrified response to being abandoned in the universe and to a fear of the oncoming darkness of death, basing this on Ecclesiastes, Job, and the Psalms. He also used Isaiah on the problem of being good. He stressed the here-and-now since "we can't help a dead man."[187,p11]

As a fifteen-year-old, Brecht wrote a play entitled *The Bible*. In it, a Catholic army besieges a Protestant town that can be saved only if the mayor's daughter sacrifices herself. Her religious grandfather dogmatically argues that a soul is more valuable than anything. However, the girl's brother argues that her sacrifice will save thousands.

Much of Brecht's early writing concerned this search for the greater morality behind middle class conventions. Brecht found the easiest way to do this was by using Swabian negativity and attacking false morality.

The Critic

Brecht's first published writing, in 1913, criticized the actions of the Great Powers towards the smaller countries then fighting in the Balkans. The powerful nations were not motivated by their claimed moral principles but by self-interest. Brecht often emphasized the Swabian dislike of hypocrisy, especially the hypocrisy of the middle class. Brecht once told a girlfriend he had to become famous to show people what they were really like. Brecht was beginning to reveal his lifelong passion to foster a better world. This would later become more overtly political.

Around 1920, Brecht wrote several one-act plays attacking the moral posturing of the middle class. In one, a strict father worries about his daughter's reputation. When her boyfriend visits, the father chases him away and beats her. Later that night, the boyfriend climbs a ladder to her window and easily seduces her. Discovered, the lovers escape onto the roof. The angry father removes the ladder, stranding them there. When the townspeople, including the pastor, see them on the roof, the father becomes the object of the ridicule he feared.

An emerging theme for Brecht was sympathy for the working class. Already in high school, Brecht cultivated a carelessness in dress and hygiene that gave him a tough appearance. He also started wearing his trademark worker's cap. His maternal grandmother's Bible readings may have helped foster his egalitarianism with its stories of how Jesus loved the poor and despised the money changers, high priests, government officials, and rich people. In fact, Brecht's maternal grandfather came from a long line of pail makers and weavers, lowly rural craftsmen.

At the beginning of World War I, the teenaged Brecht at first mouthed the patriotism he heard in school—how all Germans were ready to give their lives for country and Kaiser. But he soon wrote a poem and an essay asserting such sentiments were pure propaganda. For that, he was nearly expelled. This writing, too, could have sprung from the New Testament. "Blessed are the peacemakers for they shall be called the children of God."

When Brecht joined his intellectual friends, they talked about life, literature, and music, not just about women. Inspired by his ravenous reading, Brecht wrote of wild sex, desperadoes, and pirates. He "challenged every traditional idea of propriety" and had a "rebellious and obstreperous nature." Brecht also experimented with different attitudes and postures towards others. He was finding himself.[183,pp2,16,37]

Brecht's best friend Hanns Otto Münsterer recalled: "Where it was a matter of pure intellect and logical deduction he was often surpassed by his friends. Nevertheless, he always stuck to his guns…until the very next day, when he would expound our own opinions, as if they stood to reason and he had never doubted them."[183,p77]

Brecht already had personal magnetism. Münsterer remembered

how Brecht almost daily surprised his young friends with new poetry, some worthy, some rejected. Brecht created ceaselessly: ballads, erotic novels, brilliant maxims, scenarios and librettos for operas and oratorios. He would accept ideas and suggestions for changes, and he would take someone else's poem, seemingly a failure, and use it as the starting point for one of his own poems. These were the first signs of the writing cooperatives that Brecht later developed.[188] Typescripts later available from his writing, including poetry, show much evidence of his corrections. Almost always his changes were improvements. Trial and error would become a basic symptom of his great intrinsic motivation for writing. Failures do not decrease such motivation.[189]

To Munich and Revolution

To evade the draft for World War I, Brecht enrolled to study medicine at Munich's university in October 1917. He also attended a seminar by an eminent theater historian, hoping his assigned papers would win him entrance to the literary world. A fellow student recalled that Brecht was "self-willed and obstinate, resolute in his disregard of both praise and criticism, and self-opinionated to the point of arrogance." Brecht once criticized a classmate's play, saying it "stinks of stage-fright and stale sweat." A student felt that compared to "the run of the mill academic seminar paper, [Brecht] was a veritable grenade in our midst."[183,p159]

The military draft called up Brecht's younger brother, Walter, and then even medical students like Bertolt. According to his girlfriend, Brecht drank black coffee to overload his weak heart, but it didn't get him exempted. He was conscripted only a month before the war ended, and he served in his hometown of Augsburg at an army hospital for troops with sexually transmitted diseases.

With Germany's defeat, violence broke out in Berlin and elsewhere in the country. Navy mutinies triggered rebellions in many cities. People demanded the Kaiser step down. The new Social Democratic government of the Weimar Republic tried to keep order.

It was attacked from the left by Communists trying to copy the recent Bolshevik takeover in Czarist Russia and from the right by the military trying to destroy the democratic government.

In Munich, Kurt Eisner and members of the Independent Social Democratic Party captured military headquarters and deposed the King of Bavaria. Brecht briefly joined this pacifist party that had earlier splintered from the mainline and war-supporting Socialists. In the new Bavarian Republic, the Independent Social Democrats and the Social Democrats cooperated. They set up workers' and soldiers' soviets, or councils, to elect representatives to Bavaria's parliament. Brecht briefly represented the medical staff of his Augsburg hospital for the Independent Social Democrats.

Turmoil continued in Berlin and elsewhere in Germany. Friedrich Ebert, the Social Democratic head of the new republic distrusted the newly elected governing councils, suspecting infiltration by Communists. He turned to the political right, the army and the demobilized soldiers in the *Freicorps* (volunteer militia). In January 1919, Karl Liebknecht and Rosa Luxemburg, led a Spartacist (Communist) coup in Berlin. The *Freicorps* brutally murdered Liebknecht and Luxemburg.

Luxemburg was Jewish as was Eisner and a few of the most visible revolutionary leaders. This confirmed feelings by right-wingers that Germany might be taken over by Jewish Bolsheviks. Eisner was eventually assassinated in Munich by a right-wing law student.

Brecht was released from army service on January 9, 1919. At Independent Social Democrat meetings, he took detailed notes but never said anything himself. His girlfriend, Paula Bahnholzer, was pregnant and he needed money, so he concentrated on writing rather than activism.

From mid-January to mid-February, he wrote a major play, *Spartacus*. The hero, like Brecht, wants to change the world yet leaves the barricades to be with his girlfriend. He puts his own survival above the revolution. Brecht stressed that everyone was top dog in his own skin.

Soviet Bavaria

Bavaria's moderate-left government could not cope with the dismal economic conditions that followed Germany's defeat. Radicals and Communists infiltrated the soldiers' and workers' councils and tried to proclaim a soviet republic. Then the more extremist Communists violently took over. This lasted only a few weeks as *Freicorps* and loyalist troops from all over Germany headed to Bavaria. On Easter Sunday, 1919, Augsburg fell to 2,000 counter-revolutionaries on their way to Munich, while Brecht was meeting with his best friend to work on a script.

On that year's May Day holiday in Moscow, Vladimir Lenin bragged about the working-class victories in Soviet Russia and Soviet Bavaria. At that time, loyalist troops and volunteers, including Brecht's younger brother, were encircling Munich. Soldiers stomped Communists to death and dragged their corpses through the streets. It was a conservative White Terror much worse than the earlier Communist Red Terror.

In Russia, Communists had taken advantage of the country's defeat in war to wrest control of the government from Alexander Kerensky's democratic liberals. But Lenin complained that Germans weren't ruthless enough to carry out such a revolution. They wouldn't even attack a railroad station without first buying train tickets. In fact, the earlier Communist revolution in Berlin had failed because Germans respected authority so much: When Karl Liebknecht sent armed men to take over government buildings, a junior naval officer stopped them by pointing out that the orders from Liebknecht had not been signed.

Later, Brecht apologized for this time in Bavaria. "We all suffered from a lack of political conviction; and I, in particular, also from my old inability to show enthusiasm."[190,p25] It was always hard for Brecht to be an activist and positive. Even after he became a Communist sympathizer, he was most happy when he was critical and individualistic, a self-styled "independent Independent," who just pointed out to others what they were doing wrong.[184,p117]

All along, Brecht was collecting material for his writing. His visits to lowlife bars with black market dealers and easy women yielded scenes for his much later *Threepenny Opera*. One night in the seedy Seven Hares, one of the prostitutes climbed onto a table and sang a crude song to great applause. Brecht then took out his guitar and sang. The prostitutes and the patrons paid rapt attention to his high-pitched voice and unusual rhythm. Brecht got an ovation, a man passed a hat for contributions, and he did an encore.

A contemporary recalled, "he didn't sing well, but with an overpowering passion." A fellow medical student once noted that Brecht got ninety percent of his women that way.[183,p125]

Brecht's play, *Drums in the Night*, published in 1922, came to the attention of an influential Berlin critic, especially for its earthy, direct language, so unlike the High German usually spoken on the stage. It was the language Brecht had heard in the workers' quarter of Augsburg, in the army, and in saloons.

Brecht's love life was earthy too: he was involved with several women besides Paula Bahnholzer. The singer and actress Marianne Zoff recalled their first meeting in her dressing room: "He wasn't exactly well-groomed...I watched him with a certain fascination as one gawks at an exotic animal in the zoo...I realized that I was beginning to like this Swabian bumpkin of an intruder. He talked and talked and talked."[183,p154] Their relationship lasted two years and included marriage and a son. They broke up after Zoff found Brecht in bed with Helene Weigel, an actress who would later marry and raise a family with him. He was also involved with Elisabeth Hauptmann, who after 1925 became probably the most creative and productive member of Brecht's writing team.

At the end of 1922, based on his three plays, *Baal*, *Spartacus*, and *Drums in the Night*, Brecht won the prestigious Kleist Prize for innovative new writers. Brecht probably won the prize because his plays shook up German theater mightily. In *Baal*, Brecht had intentionally depicted crude immorality, dealing with drink, women/whores, homosexuality, murder, and an emphasis on "excrement and decay."[191,p25] *Baal* gave him the notoriety needed for his first steps to success. He would later admit its shortcomings as a "glorification of naked egoism," not at all consistent with his later emphasis on fostering a better society.[192,p16]

Unfortunately, the Versailles Peace Treaty of June 1919 did not end the great hatred between France and Germany. By January 1923, Germany had defaulted on war reparations to France, which then sent troops to run the Ruhr factories and mines for their own benefit. The German government ordered boycotts, strikes, and other resistance. The French responded with shootings, mass arrests, deportations, and an economic blockade that separated the Ruhr and Rhineland from the rest of Germany. The occupation of its industrial heartland brought Germany to an economic standstill.

The inflation in Germany that had at first developed from excessive war spending now exploded. In the first half of 1922, 320 German marks equaled $1, but by December the value of the mark fell to 8,000 to the dollar. In 1923, prices began doubling every two days. By December, it took 4.2 trillion marks to equal $1. A large laundry basket full of marks would not buy a loaf of bread. People burned valueless banknotes instead of firewood. The government had to issue increasingly larger banknotes. The largest was for 100,000,000,000,000 marks. However, the bank couldn't issue banknotes fast enough to keep up with inflation, so barter replaced cash, food riots broke out, and people with savings accounts, especially the middle class, lost everything.[193, 194]

Hitler Appears

The national crisis helped right-wing groups gain popular support, especially in Munich, which had become dominated by reactionary politics. A former Austrian corporal, Adolf Hitler, had arrived there in 1919. He took control of the Nazi Party (National Socialist German Workers' Party) in 1921. By 1923, his uniformed Nazis often paraded and heaped abuse on Jews and leftists.

In September 1923, the German government bowed to the French occupation, and both the right and the left launched revolts in Germany. The Communists struck in Hamburg, Saxony, and Thuringia. In Munich, Hitler proclaimed a *Putsch*, or coup, in a beer hall. However, the Bavarian police blocked the Nazis' march on government buildings, killing sixteen.

Brecht avoided involvement. He did not see Hitler as a real threat. In fact, he enjoyed the choreography of the massive Nazi demonstrations and Hitler's theatrical gestures. Brecht's closest friend joined the Nazis, but Brecht saw through the Nazis' red flag. They were not reds. The Nazis were a middle-class, anti-revolutionary movement. Brecht created the term "Mahoganny" for the masses of brown-shirted marchers. He announced, "Should Mahoganny come, I go."[195,p130]

Hitler's public trial for treason after his beer hall coup made his views known across Germany. He got a sentence of five years and served only nine months. During his comfortable confinement, he wrote *Mein Kampf* (My Struggle), detailing his ambitions. Such leniency for right-wing rebels was common in Germany at the time, as judges saw the Nazis as middle-class advocates for more law and order. Leftists who wanted to overturn the government, including its privileged judiciary, didn't get the same soft treatment. Brecht's plays often included courtroom scenes with corrupt judges supporting middle-class values and acting unjustly.

During the social, political, and economic disorder of the 1920s, many Germans gravitated towards extreme political parties that proposed simplistic solutions to the nation's serious problems. Moderate politicians tried to cobble together solutions but were ineffective. There was so much bickering among the various parties that President Paul von Hindenburg had to invoke a clause in the Constitution that authorized him to suspend civil liberties and rule by decree. With typical German angst many people came to prefer the enforced order and promised effectiveness of either the Nazi Party or the Communist Party.

Perhaps if Hitler's 1923 Munich coup had not been put down, it might have developed into a potent nationwide Nazi movement. There was a population bulge of rebellious youths joining the Nazi brown shirts during this time.[196]

The great appeal of the Nazis and Hitler was due to their recognition that Germans continued to smart from the defeat in World War I. People had gone from feeling pride in their country to great shame.[197] Because of President Woodrow Wilson's Fourteen Points, Germans expected reconciliation. Instead, the Treaty of Versailles took away large parts of German territory, seized all German colonies, and excluded Germany from the League of Nations as being unworthy. The Allies spoke in general terms of disarmament, but only Germany had to disarm. Almost all Germans felt the Versailles Treaty was a brutal and unfair piece of trickery forced on the German people.

At war's end, Germany had been threatened with a continued blockade on food and other necessities if they didn't accept sole guilt for World War I. The Allies enforced the blockade for another ten months after the cease-fire, causing starvation and much sense of injustice and betrayal.

While there were economic causes for the rise of the Nazis, Hitler cleverly stressed restoring pride and self-confidence in place of the humiliations at the end of World War I. This was undoubtedly why the Nazis appealed so effectively to Germans of all social classes and callings.[148, 197]

The Appeal of Communism

Brecht may have gotten involved with Communism in part because of his search for certainty. Its discipline brought order to his chaotic life full of experimentation and revolt. The Communists offered ideals, organization, purpose in life, and an intellectual framework that explained and justified much of the way Brecht already thought about society.[181, 198]

Brecht had become aware of how social forces drastically affected individuals. After he moved to Berlin in 1924, a city thirty times larger than Augsburg, he witnessed the harshness of strangers from all over Germany fighting for success. Brecht was no longer a carefree youth, but a writer trying to make a name for himself. Upton Sinclair's *The Jungle* inspired Brecht to write a play set in Chicago, *In the Jungle of the Cities*. Communism helped Brecht blame urban problems on capitalism.

From 1920 to 1925, Brecht also worked on a comedy, *Man Is Man*, showing how social forces can be harnessed to change people. Behavioristic psychology was becoming popular, and Brecht accepted its idea that external forces could transform personality. In the play, a docile man is forced into the army to become a ruthless fighting machine.

In the same way, Brecht felt that people could be manipulated to become friendly and cooperative just as Communism optimistically believed that with the right environment and the submersion of the ego, a new man could be created. This was an important part of Soviet ideology.

Brecht started reading Karl Marx, probably because of his relationship with Helene Weigel, an avowed Communist. He wanted to understand economics. With farmers worldwide going bankrupt in the 1920s, Brecht wanted to know how commodity exchanges could control prices.

Marxism was also compatible with Brecht's Swabian negativity. There was much to criticize in Weimar Germany: the newly rich lording it over others, the stock exchange gamblers, and the black marketers. In Communist theory, Brecht found a clear scientific explanation of why the wealthy oppressed those beneath them and how a greater morality would eventually triumph over the degradation in capitalistic Weimar Germany.

The self-corrective system in Marxism was also appealing because Brecht always had a firm belief in change. Marx, like Brecht, also valued materialism over idealism and spoke of the need to test ideas (praxis) rather than merely philosophizing. In addition, Marx's theory of dialectics viewed negativity as necessary to bring about change. Brecht embraced Marx because he had been thinking that way all along.

As he studied Marx, Brecht wrote another play, *The Rise and Fall of the City of Mahagonny*, depicting the many evil ways that money affects people under capitalism. Everything was permitted in the name of money. When the play opened in 1929, Nazis demonstrated in front of the opera house.[180]

Brecht made friends with several Marxists who helped him think more positively and focus his attention on the lower class. The *Threepenny Opera* of 1928 included the famous response to the fat gentlemen who criticized workers as sinful: "*Erst kommt das Fressen, dann kommt die Moral*" (First fill our bellies, then you can preach at us). *Fressen*, the word Germans use to refer to eating by animals, was a typical example of Brecht's shocking use of new language on the stage. Despite its many Marxist messages, the play was a great commercial success

Committed to Communism

Perhaps the biggest trigger for Brecht's conversion to Communism occurred during the 1929 May Day demonstrations by workers. They had been forbidden to demonstrate by Berlin's chief of police, a Social Democrat.[190] Brecht watched the May Day celebrations from his third-floor window. When he heard shooting and saw marching workers being killed, he turned white. Thirty-five were killed by the police controlled by the Socialists.[190,p149]

Pacifism had been central to Brecht's life. He could not even kill houseflies. His wife had to do it.

In 1930, Brecht began writing his didactic, or teaching-learning, plays. *The Measures Taken* tells of a Communist Party member tried for neglect of duty. He accepts that his comrades must put him to death for the good of the Party. In another play, a Chinese coolie is killed because his master mistakes an act of kindness for an attack. Brecht felt that people were naturally good but that economic competition and conflict poisoned human relations.

Brecht's humorless didactic plays were starkly staged and meant to teach, not entertain. He was also devising a new epic storytelling style of playwriting. "Epic" did not mean heroic or grand, but storytelling using dramatic means. One way to think of this approach would be as someone telling of witnessing an accident, using gestures and speaking straight to the audience.

Writing his didactic plays probably helped Brecht firm up his commitment to Communism, When people fear backsliding from a commitment, they often exaggerate.[36] Being an individualist, Brecht must have had some misgivings about accepting Party discipline. By preaching with his plays, Brecht may have been trying to firm up his conviction that Communism was right.

In this period, Brecht also wrote *Saint Joan of the Slaughterhouses*, arguing that efforts to be kind in an unkind world are ineffective. Joan, a Salvation Army worker, tries to help the miserable slaughterhouse workers. However, she draws the line at supporting their general strike. She does not believe in force.

Then, as Joan is dying, she realizes her efforts have not helped. "I have changed nothing...take heed that when you leave the world, you were not only good, but leave a good world."[195,p265] She realizes violence may sometimes be necessary to change the world. Her dying words are drowned out by the chorus of slaughterhouse owners praising her ineffectual goodness.

When Brecht completed writing the play in 1931, no theater dared produce it. The Nazi Party was so strong that most Germans feared being linked to such leftist views. It received its premier in Hamburg only in 1959, after Brecht's death. That audience gave a standing ovation for over a half hour.[199]

The Nazis Take Charge

In 1928, the Nazis had only twelve seats out of the nearly 500 in the German Reichstag. Then Hitler formed a coalition with the Nationalist Party, which held seventy-eight seats. The head of the Nationalists, Alfred Hugenberg, was a former director of Krupp industries, and he had taken advantage of the runaway inflation to buy up newspapers, magazines, movie houses, and a news film company. All of his media publicized Hitler and the Nazis. In addition, the Nazis took great pains to ingratiate themselves with "people who mattered" including top people in the military and industry. This all helped make the Nazis socially acceptable.[148,p31]

The efforts paid off handsomely. In the 1930 election, there were 4.5 million new voters, many of them idealistic youths seeking new directions for Germany. The Nazi vote increased eightfold that year, and the Nazis became the second-largest party in the Reichstag with 107 seats.[196, 200, 201]

The 1929 stock market crash in the U.S. triggered a worldwide economic crisis. The following banking collapse and the beginning of the Great Depression soon put millions out of work. In Germany, five million were unemployed by the summer of 1930, six million by the end of the year out of a working population of twenty nine million.

As in other countries, German Chancellor Hermann Brünning responded with fiscal austerity that only made the deflation worse. In 1930 wholesale prices fell 13 percent. With the financial crisis, investors feared for the security of their money. Bank withdrawals reached panic levels. In the September 1930 election, the Communists polled 4,600,000 and the Nazis 6,400,000 votes. It seemed only one of the extremist parties could restore order.[202,pp85,101]

In 1929, Erich Maria Remarque's book, *All Quiet on the Western Front*, had been a best seller in Germany and in many other countries. It attacked German militarism and stressed the terrible toll of World War I. But in December 1930, when the film version was shown in Berlin, Dr. Joseph Goebbels called the Nazi Storm Troopers into the streets. Riots and demonstrations made the authorities ban the film as an insult to the German army and the German nation.

Many who joined the Storm Troopers had been born in the few years after 1900 and politicized by war defeat, Communist revolts, and right-wing repression. Memoirs by Storm Troopers have recounted their backgrounds of poverty and the lack of fathers at home.[196] They had belonged to various youth groups that gloried in violence for its own sake. Many sought a quasi-religious and classless comradeship. The Nazis were appealing because they promised violent action under cover of a legitimating ideology. On the streets and in meeting halls, the Storm Troopers battled Communists as well as the Social Democrats who were defending the Weimar parliamentary democracy. Another set of self-descriptions written by Nazi militants in 1936 reflected similar needs for comradeship, sadism, and paranoid scapegoating.[203]

Enormous parades and rallies convinced people that the Nazi Party had much popular support. Conservatives in the working and middle classes were seduced by the enthusiasm of great masses of Nazis. Many saw the Storm Troopers as the only force able to stand up to the seemingly foreign or un-German Communists. Hitler's clever strategy was to create a mass movement that transcended the rigid class divisions then existing in Germany. Membership in the Strom Troopers grew thirteen fold in the years just before the election of 1932.[196, 204]

In that election, the Nazis got seventeen times more votes than in 1928. They promised strength to end the confusion, indecisiveness, and quarreling of the Weimar Republic's political parties. Germans wanted a powerful government to rule and reverse the humiliating Versailles Treaty. The Nazis had 230 seats after the 1932 elections and held the balance of power in the Reichstag. It wasn't the prospect of solutions to the serious economic and political difficulties, but rather Hitler's promise to avenge the Versailles Treaty, that made him so popular across all segments of German society.[148]

The military and wealthy industrialists who held the real power in Germany now had to deal with the Nazi Party. However, it was too late for them to stop the Nazis.

Hitler and conservative ex-Chancellor Franz von Papen convinced German President von Hindenburg that a leftist coup was impending, and on January 30, 1933, Hitler was appointed

Chancellor, head of the government. In Berlin, strangers embraced, feeling their savior had arrived.[205,p232]

Von Papen became Vice-Chancellor and Premier of Prussia, but Hitler cleverly appointed henchman Hermann Goering as the Minister of the Interior for Prussia. Goering thus controlled the police in the two-thirds of Germany that was Prussia. Four weeks later, the Reichstag building burned in a fire blamed on a Dutch Communist. That night, Goering ordered 4000 Communists and other political opponents of the Nazis arrested for complicity in setting the fire.[206,p27] Few escaped. During 1933 and 1934, more than 100,000 experienced the concentration camps.[145,p42]

Brecht escaped arrest because he was not at home but sleeping at a medical clinic. He fled the next day to Czechoslovakia. Brecht was reputedly Number Five on Hitler's death list.[191,p47]

That same day, Hitler also had President von Hindenburg suspend civil liberties in favor of rule by decree. A national election a few days later gave the Nazis enough votes to form their own governing coalition. But that was merely a formality because the Nazi police state was already in power. Two months after Brecht fled, the Nazis burned his books together with those of Einstein and other leftists.

Into Exile

Perhaps the biggest sacrifice Brecht had to make for his commitment to bettering the world was his fourteen years in exile where he no longer had German audiences or actors. Of course, even before he left Germany Brecht's commitment had been tested in many ways. There were fascist demonstrations outside his plays and eventually, as the Nazis became stronger, complete refusal by theaters to put on his plays.

A major sacrifice in America was not being able to write films and plays in his usual epic and critical fashion. Of the fifty film scripts he wrote, he only got some money for two. He was also rejected on Broadway. His leftist reputation got him called before a

Congressional committee that hounded him out of America.

Even after World War II, Brecht found the dictatorial control by the East German government prevented him from writing the plays he wanted. However, none of these problems could prevent him from continuing to try. Nothing would persuade him to give up his basic commitment to justice for those at the bottom of the heap.

Brecht lived in Scandinavia from 1933 to 1941, first in a house in Denmark scarcely forty miles across the water from Germany. He told another refugee, "Don't go too far away. In five years we shall be back."[195,p292]

Exile is especially tough for all writers because they no longer have audiences who understand the writer's language. As a playwright, Brecht also lacked a theater, actors, and directors. He realized "teaching without pupils—writing without glory—is difficult."[195,p296] Fortunately, he viewed his situation with humor. He kept writing "for the drawer," while some eminent exiled German intellectuals even committed suicide.[207,p34]

Brecht applied for an American quota visa for himself and his family. Unfortunately, after the Nazis stripped him of his German citizenship, he no longer qualified for the German immigrant quota. He was placed on a much longer list for those without nationality.[207]

As Hitler's preparations for war became more ominous, Brecht moved to Sweden. The Germans invaded Denmark and Norway in April 1940, and then the neutral Swedes grew nervous about offending Germany. After Swedish police looked into Brecht's anti-Nazi activities, he moved to Finland. That country also became untenable when Germany pressured the Finns to accept German troops to defend them against the Soviet Union.

Eventually Brecht got help from friends in America. He got an invitation to teach at New York's New School for Social Research, qualifying him for a non-quota visa. This prompted him to go the American embassy in Helsinki to re-enter the names of his family and of collaborator Margarete Steffin on the waiting list. Brecht had not thought to do this because he felt their situation in Europe wasn't yet hopeless.

Brecht could not leave without Steffin. As he wrote to the New School, "She...is the only one who has an overview of my thousands of manuscript pages. Without her, preparing lectures would mean an enormous loss of time. For ten years she has been my closest collaborator and is personally much too close."

A day after the visa for Steffin arrived, they boarded a train for Moscow on their way to Vladivostok and a ship for America. While the ship was in the Pacific, the U.S. Government declared that people with close relatives in Germany, like Brecht, could no longer apply to enter the U.S. This was one more close call that Brecht evaded.[207,p24]

Brecht's writing during his Scandinavian exile continued to stress moral issues. *The Fear and Misery of the Third Reich* contains twenty-eight poignant scenes. (Also titled *The Private Life of the Master Race* in a shorter English version.) In one scene, a fisherman on his deathbed asks a pastor if Jesus was right to say in the Sermon on the Mount that peacemakers are blessed. Or were the Nazis right, since Jesus was just a babbling Jew? The pastor responds with the traditional Lutheran view that the church should not get involved in politics.

In an anti-war play Brecht wrote in Sweden in 1939, *The Trial of Lucullus*, a jury meets at a Roman general's grave to decide whether he should go to heaven or hell. Did he accomplish anything worth the deaths of 80,000 soldiers? The play emphasizes the futility of war.

Another pacifist play, *Mother Courage and Her Children*, stressed the mercenary character of war. It is based on Grimmelshausen's seventeenth-century story of a woman during the Thirty Year's War in Germany. The mother loses her children one by one in the war, yet she continues selling supplies to the soldiers. Brecht's point was that not only big profiteers fuel conflicts, so do ordinary citizens, even those who lose their sons.

Brecht referred to his long period of exile as "changing countries more often than shoes."[179,p62] Why didn't Brecht take refuge in Moscow, as had Walter Ulbricht, later to become head of Communist

East Germany? Apparently he was not completely committed to the Party line. He had visited Moscow briefly in 1935, learning that his views on art conflicted with the Stalinist requirement for socialist realism. According to biographer Eric Bentley, Brecht "saw Russia as on the right path, not by any means as having arrived at the goal."[208,p170]

Supposedly, Brecht at various points did defend Stalinism, including the 1930s show trials and the killings of his friends there. Apparently he told noted anti-Communist Sidney Hook, "The more innocent they are, the more they deserve to be shot."[209,9493] He also said, "With total clarity the trials have proved the existence of active conspiracies against the regime." Brecht's statements were evidence that he was a Stalinist—or was he pulling Hook's leg?

Hollywood

Arriving in America after eight years of exile, Brecht was financially strapped. He decided to write for Hollywood rather than going east to take a chance on Broadway. Living costs would be considerably higher in New York City, and he had his family to think about: his wife, son, daughter, and his mistress-collaborator, Danish actress Ruth Berlau. Margarete Steffin, with only one lung, had died of tuberculosis in Moscow.

Brecht had been deeply shaken by Steffin's death, and the move to America disoriented him even more. Nine days after arriving, he wrote in his journal, "It's as though they'd taken away my guide just as I entered the desert." In addition, he felt the loss of the amateur actors and political partners he'd had in Europe. Nine months after arriving in America, Brecht realized that for the very first time he was no longer doing proper work.[207,p33]

Hollywood was a major shock for Brecht. All his life he had a Swabian commitment to living simply and frugally. Now he was in the lush semi-tropics, so different from Swabia and from Scandinavia. He wrote a poem about Los Angeles as hell, a place with luxuriant gardens that wilt without expensive water, heaps of fruit without smell or taste, and fine houses built for people who were unhappy because they couldn't pay the bills.

Most of all, the commercialism of Hollywood conflicted with Brecht's values, confirming his earlier view that everything was a commodity in capitalist America. In January 1942, he wrote in his work journal how he often looked for little price tags on hills or lemon trees, on people's ideas, and even on their gestures.

Brecht wrote critical poetry about the movie moguls who bought lies (not objective truth) and wanted to be told how great they were. When he looked into their rotting faces, he could no longer eat. Brecht felt he could still smell the stench of the moguls' moral decay even where he lived in Santa Monica, far from Hollywood.

Brecht knew he would have to make some compromises in his film writing, but he thought he could keep his basic integrity. Even though the industry wanted patriotic war movies, he continued to write pacifist scripts. Hollywood wanted entertainment, but Brecht wanted to raise consciousness. Only two of the fifty scripts he worked on were sold. He wasn't even listed in the credits for *Hangmen also Die*, since his work was mostly deleted, and he only supplied the idea for the second film, *Simone*, which was never produced.[207]

However, just the money from the first film bought him enough time to write three plays. His response to the second script's sale was typical of a frugal Swabian. He used the money to pay off his home mortgage, put the rest in the bank, and continued to live just above the poverty level. One observer described him as living "an almost Gandhi-like ascetic existence, occasionally seasoned with small pleasures." Later, after he died in East Berlin, people were shocked at how very simple his apartment was.[195,p382]

For language reasons, many of Brecht's friends in Hollywood were German emigres. However, he selected friends who did not brag about the superiority of German culture. Brecht thought his own ideas about theater were superior, but not because German culture was. As befits a committed workaholic, Brecht had friends who were mostly work-related, fellow artists, writers, directors, actors, and musicians. They ranged from conservative to Marxist and often became collaborators on Brecht's writing projects.

Brecht's Enemies

Brecht also made enemies. From his youth, in his first published writing as a theater critic in Augsburg, he had been outspoken like so many Swabians. When he saw something wrong, he was not afraid to attack.

Brecht's dislike of Thomas Mann, the most famous German writer of the time, dated to the 1920s. He thought Mann was too pompous, which contrasted with the egalitarianism of a Swabia long free of a landed nobility. Thomas Mann was too good and too wise, a show-off. Mann gave marvelous impromptu speeches, apparently rehearsed in front of a mirror. This may have given Brecht the idea for writing a scene in which Hitler practices his public gestures in *The Resistible Rise of Arturo Ui*. Thomas Mann, for his part, once described Brecht as "very gifted, unfortunately."[208,p32]

Brecht also disliked the three leftist intellectuals who came from the Institute for Social Research at the University of Frankfurt: Theodor Adorno, Max Horkheimer, and Herbert Marcuse. He may have been influenced in part by the Swabian peasantry's traditional dislike of the Württemberg regime's educated "scribblers" who did nothing but take their money.[210,p31] Brecht thought these intellectuals were not independent freethinkers but men who created and sold ideas at the bidding of others. Marcuse did work for the U.S. Office of Strategic Services and later its successor, the Central Intelligence Agency.

These Frankfurt intellectuals claimed to be Marxists, but instead of applying their ideas to the real world (what Marx called "praxis"), they simply spun theories like many other Germans. Brecht disliked intellectualism without action and thought these men were pontificating elitists who, like Hitler, believed German culture was superior. They also lived like other well-to-do people. Brecht expected committed revolutionaries to live modestly, not like capitalists. The endowment for the Frankfurt School's Marxist studies had originally come from a wheat speculator in Argentina.

In all these ways, the Frankfurt intellectuals were the opposite of Brecht. For their part, they saw him as a simplistic, "vulgar" materialist who preferred the proletariat to progressive-minded intellectuals like themselves who would launch the revolution.

Failure on Broadway

While Brecht was working without much success in Hollywood, he really wanted to succeed on Broadway. In 1943, he attended an antifascist rally in New York City that included a short sketch written by two Czechs and based on Jaroslav Hasek's *Good Soldier Schweyk*. This inspired him to write *Schweyk in the Second World War*, a comedy. Brecht knew a tragedy would not succeed on wartime Broadway.

Unfortunately, Brecht's characterization of Schweyk, a peasant type that Europeans would find funny, was strange to Americans. And Brecht's vulgar anal humor in this and other plays was more fitting for a German audience.[211] Finally, the basic premise of the play—that a widespread peasant-like emphasis on survival can sabotage a regime—seemed bizarre to Americans. Brecht idealized the common people, disregarding that many of them had been enthusiastic for Hitler and the Nazis. (Only decades after World War II did Germany acknowledge anti-Nazi rebels such as those in involved in the White Rose conspiracy or von Stauffenberg's Valkyrie coup and attempt on Hitler's life.)

Brecht also revised an earlier work, *The Caucasian Chalk Circle*, for Broadway. It is based on an incident during Brecht's childhood when someone stole pears from the family's tree. Brecht told his father not to get upset because the thief probably needed the fruit more than they did.[212]

In the play, a provincial governor has to abandon his child during a revolt, but the maid rescues and raises the boy. Eventually, the governor's wife goes to court to try to reclaim the child. The judge is a drunken village scribe who, like so many of Brecht's judges, is corrupt. But his humble background gives him sympathy for the downtrodden. He draws a chalk circle around the child and has the two women stage a tug-of-war over the child. The maid sees the child hurting and lets the mother win. But because the maid clearly loves and needs the child more, the judge awards her the child.

In a concession to American audiences, Brecht substituted a happy ending for his usual open, thought-provoking ending. The play still could not get produced on Broadway. Only in 1948 was it performed by a college group in Minnesota.

Starting in late 1944, Brecht worked with the British actor Charles Laughton to translate and revise a 1938 German script about Galileo into English. Brecht liked Laughton because he had a tireless work ethic, was not pretentious, disliked authorities, and sympathized with the common people. They spent more than three and a half years working together. Every day for two of those years, they acted out passages with gestures and searched for the proper English words. They focused on the behaviors of the characters rather than their psychology. Brecht wanted gestures and music to express the attitudes of speakers.

In a 1940 poem, Brecht had described learned scholars as butchers who emerged from libraries while mothers and children searched the sky for their horrible inventions. The new version of *Galileo* included references to the 1945 dropping of atomic bombs at Hiroshima and Nagasaki and stressed the need for social responsibility by scientists. After a successful Hollywood production of *Galileo* in the summer of 1947, a Broadway production was scheduled for later that year, but by then Brecht had left the United States for political reasons.

Political Problems

For some time, the FBI had monitored Brecht and his wife. With thousands of other German aliens, they had to stay close to home except to commute to work, and they had to observe a curfew from 8 p.m. to 6 a.m. Brecht thought these restrictions were reasonable. Even before stepping foot in America, he had tried to avoid problems by throwing his books on Lenin overboard into Long Beach Harbor.

By 1943 the FBI had transferred Brecht's dossier from "enemy alien control" to the more serious "internal security control" for Communists. The government's surveillance was so obvious that his wife once told an agent in a car parked outside, "You poor man, you could watch us much better from inside the house."[207,pp314-5] Brecht's frequent criticism of others had also made people happy to inform on his activities.

As the Cold War with the Soviet Union intensified, the infamous House Un-American Activities Committee (HUAC) subpoenaed Brecht and eighteen other Hollywood figures suspected of being Communist agents. The subpoena warned him not to leave Washington without HUAC's permission.

At the October 1947 committee hearings, Brecht used the peasant's traditional approach of playing dumb and apparently cooperating with authorities. Brecht politely agreed to give full answers. After all, he said, he was a guest in America, not a citizen like the previous defendants who had invoked the First Amendment guaranteeing freedom of speech and belief.

Even though Brecht spoke English well, he accepted an interpreter. This gave him more time to plan his answers and to hide incriminating information. Brecht rehearsed with a friend who asked likely questions. He also planned to smoke cigars because the chairman, J. Parnell Thomas, was a notorious cigar smoker. Brecht's performance was a masterpiece of cunning and deception.[207]

There were eighty reporters, as well as photographers, newsreel cameras, and radio stations. Apparently, the taped transcription revealed tension in Brecht's voice. He probably felt the Committee might block his exit from the United States.[207,p334, 213]

Asked about his leftist writings, Brecht admitted he advocated revolution—but only against Hitler's government. As to a revolutionary song he wrote that had been translated and published by the Communist Party USA, Brecht said that had been done without his permission. He corrected his interpreter to give the proper English translation of some German words—not the dangerous "take over" but "take the lead."

Significantly, Brecht testified that he was never a member of the Communist Party. HUAC could not have proved that charge anyway without evidence from Germany twenty years earlier. Some say he probably never officially joined the Party. While he was a Marxist, he probably wasn't a Communist.[195,pp506]

The committee members complimented Brecht for being so cooperative. The next day, he fled to Europe, the airplane seat reserved under another name. Brecht would use the same peasant-like approach when he later displayed ironic politeness towards the Communist East German authorities. Brecht preferred survival and success for his work rather than striking some heroic pose.

The Communist government in Soviet-occupied East Germany indicated that they would welcome such a famous German playwright. However, Brecht first settled in Switzerland, where he could once again work with German-speaking actors. He then got an Austrian passport by promising to live there. He also signed with a West German publisher to get royalties in an easily convertible Western currency. He then waited until the East Germans offered his wife, Helene Weigel, the directorship of a subsidized theater troupe numbering 350 in East Berlin.

A Bad Fit with American Theater

Why did Brecht's playwriting fail in America? Besides language and cultural problems, his plays were too impersonal for American audiences. Brecht liked the epics of old, where the actor was mainly a storyteller, reciting and showing what the characters did. The actors in Brecht's plays would give only a limited portrayal of their characters. As for many Swabians, tangible behaviors were more important than personality traits. Brecht especially wanted audiences to see that people's behaviors could easily change under different circumstances. He didn't want audiences to see a character so well defined that people might feel "that could be me."

Brecht especially wanted the audience to think about what they saw. Brecht claimed, "I have feelings only when I have a headache—never when I am writing: for then I think."[179,p239] He did not want empathy to overcome the audience's judgment. He termed that "culinary theater," where audiences felt good, but the play had no lasting impact.[195,p201]

To stress Brecht's preference for learning as an outcome, the University of Michigan for a time hosted a Brecht theater.[214] His plays have also been favorably received at other universities.

In most of America, however, theatergoers were more satisfied with actors following the Stanislavsky System and its emphasis on inner, psychological truth.[215] Brecht's impersonal and serious plays were strange. According to Eric Bentley, Brecht's translator, American publishers were uninterested in Brecht's plays.[216,p7]

Also problematic for Americans was how Brecht incorporated multiple innovations in a play, not just one or two. He might have new lighting (all white), new costuming (real, worn clothing), new stage design (only objects part of the action), new music (commenting on and contrasting with scenes and words), as well as new acting and new playwriting.[208]

There were added problems on Broadway. There, each play had to stand on its own feet financially, unlike state-subsidized European theaters that used a repertory system. European theaters could produce some risky new plays by alternating them with proven classics. Fortunately, East Germany later subsidized his theater so lavishly that Brecht was able to spend ten months rehearsing for a run of *The Caucasian Chalk Circle*.

Of course, after Brecht later settled in East Germany, he was dismissed by Americans as a "Commie." Even before that, his emphasis on the problems in society made for depressing rather than pleasant theater, turning off Americans.

Among Germans Again

In his six years in the United States, Brecht also experienced problems stemming from the many differences between German and American cultures.

Swabians expect to give and receive criticism. Swabians are suspicious of praise. It's rare. They feel "Whoever praises me either wants something from me, or he's already gotten it from me." In contrast, Americans like people to smile, be agreeable, and act friendly, and they found Brecht's combative approach grating.[2, 7] Americans often viewed opposition as a personal insult, rather than a way of getting at truth. Charles Laughton's wife described Brecht as anti-everything and thought that if he had ever joined the Communist Party, he would have become an anti-Communist.

Brecht was also too practical for American intellectuals who accused him of not valuing knowledge as an end in itself. In this, he followed the philosophy of Marx, whose gravestone in London's Highgate Cemetery reads: "The philosophers have only interpreted the world in various ways; the point, however, is to change it." Brecht insisted that writing should be of practical use. *Brauchbar* (useful) was one of his favorite words. His plays were intended to get people to think about how to make the world better.

A close friend, Hanns Otto Münsterer, recalled how as a nineteen-year-old Brecht was already taken with the view of Swabian playwright Friedrich Schiller. Theater should have a moral impact. This was similar to the pietistic view of the Bible—that it was not so much a Holy Book as a text that gives solutions to the many problems of Christian living.

Brecht liked all of his writing to be useful, even if the purpose of some poems was just to amuse friends or to seduce women. He produced such practical poems almost daily—like his gesture of thanks "To a Woman Colleague Who Stayed Behind in the Theater during the Summer Vacation."

Brecht wanted to spur audiences to think and then act on what he depicted on stage. One technique was highlighting contradictions between what characters thought and said and what they did.

He also employed open-ended climaxes. In defiance of Aristotle's pronouncement that a play should be complete in itself, Brecht wanted dramatic tension to linger without a resolution. Brecht's *Mother Courage*, for example, ends with the heroine acting as if she has learned nothing from her wartime suffering. American audiences would rather she come to a conclusion like "war stinks."

Instead, Brecht wanted the audience to think about war after they left the theater. They should ponder how "war kills all human virtues and must therefore be opposed with every possible means."[199,p260] Brecht used the Zeigarnik effect to make tension linger until some action was taken. This is based on the observation by Bluma Zeigarnik and gestalt psychologist Kurt Lewin that Berlin waiters remembered unpaid orders but quickly forgot paid ones.[217]

Formality and Privacy

Americans also objected to Brecht's impersonal way of relating to people. Eric Bentley, Brecht's translator and frequent associate for fourteen years, accused him of being uninterested in people and their feelings. He claimed Brecht seemed unable to socialize and had no interest in developing friendships. Englishman Paul Johnson similarly claimed of Brecht, "he preferred ideas to people, there was no warmth in any of his relationships," and "he had no friends in the usual sense of the word."[218,p196]

Brecht did gather people around as stimulation for his work, but he was not by nature sociable. What Bentley and Johnson did not understand was that Brecht, like many Germans, separated relaxation from work. He relaxed privately with family and close friends. At home he could relax and horse around with his family.[180,p125]

The psychologist Kurt Lewin felt German personalities were like an onion.[219] Private selves were the innermost layer and protected from other people. In contrast, Americans were more likely to "wear their hearts on their sleeves" and to be "sincere."

This German distinction between public and private is reflected in the language, which has formal and informal terms of address. Moving from the formal word for "you," *Sie*, to the informal term, *Du*, used for a friend or intimate, often involves an agreement, sometimes even a little ceremony. Germans who visit the United States can complain that since Americans seem friendly to everyone, their friendliness has no value. Probably Americans saw Brecht as unfriendly also because of his typical Swabian frankness and contentiousness.

Germans prefer to be serious in public; they value self-control. That doesn't necessarily mean they ignore feelings. People with a pietistic bent, for example, feel an obligation to examine their inner feelings. Brecht often expressed such feelings in his poetry. One evening, he talked with his best friend's wife about how they had lost so many friends during the Holocaust. The next morning she found a poem, "The Survivor," slipped under her door. It expressed the guilt and self-hate many survivors feel.

Brecht's humor was also mostly unappreciated by Americans. It was often "dry" intellectual humor. Brecht once asserted that Hegel's book on logic and dialectics was one of the greatest humorous works of world literature.

Still another American criticism was Brecht's lack of personal hygiene. He wore the same clothes for weeks at a time. He didn't bathe. Like other Swabians with great intrinsic motivation, Brecht lived simply, focusing on his calling in life. He cared little for personal appearance and fashion.

Einstein's attitude was similar. Intrinsic motivation for his scientific work trumped personal appearance and cleanliness. He wrote his second wife on December 2, 1913, during the first years of their courtship: "If I were to start taking care of my grooming, I would no longer be my own self...So, to hell with it. If you find me so unappetizing, then look for a friend who is more palatable to female tastes." [104,p58] After she died, Einstein gave up regular bathing.

Americans also criticized Brecht for sexism. Brecht's cavalier attitude appears in a 1921 diary entry about two copulating dogs unable to separate. They caused an uproar in the street as people tried to pull them apart. Brecht felt the female dog was fulfilling her duty to keep still and abandon herself to the male. On another occasion, a friend doing research for a play expressed shock that prostitutes in busy Italian seaports had to service up to 100 men a night. Brecht dryly replied the limit should be eighty.

Brecht and His Women

Despite his occasional disparaging comments towards women, Brecht felt dependent on them. "The Song of Sexual Dependency" in *The Threepenny Opera* was so explicit that it was censored in Germany for a long time. In the song, Macheath admits he can be easily seduced by any woman he allows close. Even while he flees the hangman, Macheath thinks about girls. In Macheath's song, many men turn to the Bible, to anarchism, or whatever else, to escape this sexual dependency. However, by nighttime they end with the whores. From the evidence in Brecht's many letters to women, he seems to have had this dependence on women, too.

While Brecht had long relationships with several different women, he also had briefer mistresses, often with more than one at the same time. He would get involved with an actress as they developed a new epic role. Such women gave him creative energy, love, and companionship. Brecht asserted the revolution would create free, loving partnerships between men and women as well as more caring between people in general.[220, pp219-20]

Perhaps Brecht's emotional need for these women equaled their need for him. Work and love were intertwined. However, his wife Helene Weigel once told their daughter, "Your father was a very faithful man. Unfortunately, to too many."[221,p15]

A major controversy has developed over whether the women in Brecht's harem wrote his plays. Each of the four important women—his wife Weigel, Elisabeth Hauptmann, Ruth Berlau, and Margarete Steffin—contributed in major ways to his work. They researched, translated, wrote, edited, and provided many creative ideas. The evidence is clear. However, influence must have also been in the other direction, as his helpers wrote in the Brechtian manner.[208] Male colleagues collaborated with Brecht, too.[183,103]

Brecht and his collaborators also borrowed from others' writings, and the resulting plays were often a patchwork from different sources, such as K. L. Ammer's *Villon*. When Ammer complained, he was given a share of the royalties. Lotte Lenya, star of *The Threepenny Opera*, recalled a friend saying, "Why deny that Brecht steals? But—he steals with genius."[222,pv]

Collaboration was always a part of Brecht's work. A colleague, Bernhard Reich, said Brecht would test visitors by reading a part, get reactions from them, and then immediately hammer out a new version on the typewriter. "He understood that the work profits if many take part in it."[213,p6]

Marthe Feuchtwanger, the wife of Brecht's mentor, recalled that Brecht "always had to have an audience. He would be inspired in his work when people would express other opinions and give him new lines and new outlooks."[213,p73] Weigel often put on Sunday night "kindergarten" socials at their home, so Brecht could be surrounded by stimulating friends and collaborators.[207,p227] Such collaboration was a bit like the workshops of Renaissance painters who produced works signed by one artist that were the work of many.

Brecht's charisma and genius made it easy to attract collaborators. Ruth Berlau observed: "the duties were so interesting that one was glad to take them on voluntarily. Brecht never forced, or even asked, anyone to do anything at all for him. Never."[223,pp3-4] On the other hand, Brecht's neglect landed Berlau in a mental hospital, but then he made a special trip to the East Coast to get her out and help her recuperate.

Brecht's female collaborators worked with him during the time when many women found it difficult to publish or to direct plays. The women were dedicated leftists who believed in Brecht's criticism of capitalism. Even some modern German feminists have defended the collaboration, given that women had few other opportunities at the time.[221, 224]

Some of the fiercest criticism of Brecht's use of women has appeared in John Fuegi's book that compares him to Hitler and Stalin and claims he was a psychopath.[225] Brecht scholars have attempted to answer with a chapter-by-chapter refutation of the book, however often in the same emotional style of character assassination. The scholars on both sides seem to have ignored Brecht's emphasis on the dialectic and on open discussion to determine the truth.[226].

It is true that Brecht had collaborators for his plays. Clearly, however, most of Brecht's more than 1,000 poems were his alone.[182] Besides, these poems informed most of his plays. Even though the plays were written in prose, they can often be considered a "full-scale stage poem".[189,p89]

A second point in Fuegi's critical book is that Brecht masqueraded as poor but was not. This seems true. It was a common Swabian practice drawn from Calvinism: "to have it, but not show it."

Most important, Brecht felt he was not just a playwright. His life should be seen in its totality. Early on, "he was determined that his own life should become his most important work of art, greater than all his literature."[183,p46] His prime criterion was that his activities be useful moral critiques of capitalism and of war. That was true whether he was writing or directing plays, writing poetry, or supporting others who shared his fervor for social change.

Tinkering in Berlin

Brecht especially valued working in East Berlin where he was allowed time to do much tinkering with productions, altering and fixing parts as rehearsals continued, a prime aspect of his intrinsic motivation. He enjoyed adjusting the text, lighting, music, costumes, props, and staging.

According to Brecht's longtime mentor, Lion Feuchtwanger, Brecht was "more interested in the work than the finished result; more in the problem than the solution; more in the road than the goal." He would rewrite twenty or thirty times and once more for each provincial production. He was not interested in a work being complete. He clearly had intrinsic motivation.[227,p19]

Even when an audience liked a play, it was not finished. When Brecht directed, every performance of a play was different. He was always changing one bit or another.

The Dialectic

Dialectical thinking, the contrasting and combining of opposites, developed by the Swabian philosopher Hegel, is often an important source of creativity.[108] Swabian peasants were notorious for their negativity. For Brecht, such negativity was a catalyst. Old ideas needed to be challenged to give birth to new thoughts. He valued the way in which criticism can force a person to rethink arguments and come up with better ones.

Brecht had used negative thinking in crafting his first full-length play, *Baal*. He wrote the opposite of everything his instructor wanted. Instead of idealistic complexity and abstraction, Brecht wrote scenes that were materialistic, direct, and concrete. Instead of a romantic who rhapsodizes about his love for a sweet woman, Brecht had Baal feel her warm hips in his hands. When she becomes weak in his presence, it is because she smells him.

In East Berlin, Brecht avoided psychological discussions with his actors. "Why give me reasons?" he would say to them. "Show me."[195,p461] Often, he would first have an actor perform a gesture by empathizing with the character, feeling exactly the emotions that Brecht did *not* want to guide their performance. Assistants and advisors were also asked to play devil's advocates in helping Brecht decide on a particular staging.

A year before his premature death, Brecht proposed that his plays be no longer labeled "epic theater" but "dialectic theater."[208,p259] Two phrases Brecht used in rehearsals show this change and how nothing he had written was sacrosanct. "I'm not trying to show that I'm in the right, but to find out whether." Another time he said to an actor, "To hell with the play, it's your turn now." If needed, whole new speeches could be inserted, new characters invented, even whole scenes created to fix a problem in a production.

"Dialectic theater" meant that Brecht's work was constantly in flux.[228,p154] This is why he characterized himself more as a Marxist, less a Communist. He refused to accept anything as fixed. He was always questioning.

Communist East Germany

Brecht's moralistic commitment to helping the downtrodden, around which almost everything in his life revolved, helps explain why he persisted with Communism after the Nazis were defeated. Communism claimed to make life better for the weak. It would bring about a world where people would be good to one another. One of Brecht's poems described his vision of the coming revolution in which government officials were getting friendly and the landlord visited tenants to check on the plumbing.

To Brecht, another important aspect of Marxism was that it seemed rational and self-correcting. During the Weimar Republic, Brecht was, like other Germans, faced with a choice between Communism and Nazism. The latter was based on a mystical faith in the German people and the charismatic Hitler's interpreting their will. Brecht preferred Communism because, according to Marx, it considered empirical evidence and used praxis to test ideas and to direct change.

During Brecht's postwar years in Communist East Germany, he stubbornly continued to believe in these principles even though the leaders were betraying the revolution. Pacifism also continued to be key for him. When Soviet tanks put down the workers' uprising on June 17, 1953, Brecht sent a critical telegram to Party Secretary Walter Ulbricht, asking for a dialogue between the government and the people.

However, according to Brecht, the Party newspaper printed only the telegram's last sentence, his token expression of support. The full letter was never made public. Unfortunately, his words of support discredited him in the eyes of those few .people who had not already written him off for his postwar settling in Communist East Germany.

Brecht's later unpublished poem, "The Solution," noted how the regime had issued leaflets attacking those who revolted. Since the government no longer had faith in its citizens, Brecht sarcastically suggested the government should just elect new citizens.

It's unclear whether it was a good idea for Brecht to move to Communist East Germany after he was unsuccessful in finding a permanent theater in Zurich or Vienna. Apparently, Brecht didn't have any other good alternative after the negative American response to his attempt to settle in Munich.[191] In any event, he did get his plays produced. He needed that, since he did not write finished plays.

Not everyone saw his going to East Germany as a mistake. Englishman Paul Johnson, in his book *Intellectuals*, wrote that "in the two decades after his death, the 1960s and 1970s, he was probably the most influential writer in the world."[218,p173] One American publisher reported that his books were "selling like hot cakes in every college bookstore in the land."[208,p168] This was despite the negative political attitudes that prevented an honest appreciation of his plays by many people.[191]

In Germany, he was eventually again accepted. As the *New York Times* headlined in 1998, "Beyond Rancor, Brecht Remounts his Pedestal: Playwright Is an Icon Again as Many Forgive His Marxist Politics." The *Economist* in October 2006 similarly reported on productions of *Mother Courage* on Broadway (with Meryl Streep) and with the English Touring Theatre in Britain. The Royal National Theatre was also putting on *The Life of Galileo*.

Brecht died at age fifty-eight, his weak heart finally stopping. Even in the instructions for his funeral rites, he tolerated no pompous or showy displays. He wanted no lying in state, no graveside speeches, and no epitaph. He wanted his gravestone to read simply "Bertolt Brecht."

A fitting inscription would have been some of the words from his

American poem, "Everything Changes," for they encapsulate his lifelong commitment to change and trying everything to make the world a better place. Brecht emphasized that what is done is done. One cannot separate the wine and water once mixed together, but we can always make a fresh start with our last breath.[182,p400]

7: BETWEEN HITLER AND STALIN

A common story about the farmers in Bessarabia: One morning a Bulgarian, a Romanian, a Turk, and a German heard that all the land they could pace off by the end of the day would be theirs. After an hour, the Romanian decided to stop, sit down, and eat breakfast. The Bulgarian didn't rest till his midday meal. The Turk finally halted in the evening, at supper, saying, "Anyway, that's all I need." But the German kept going and going. He was finally so exhausted, his legs gave way. But as he fell, he stretched out his hand as far as he could reach, "Up to here is mine." Then he died. Germans had very strong motivation to be farmers.

Swabians have scattered all over the world. A section of Antarctica is named New Swabia, although no one has settled there. But Swabians do have a history in Bessarabia, and I was fortunate to be able to interview a couple who used to live there.[229]

Bessarabia

At the beginning of the nineteenth century, the ancestors of Immanuel and Johanna Weiss left Swabia to take up the Russian Czar's offer of land in Bessarabia, which he had recently seized from the Turks.[230-232] Bessarabia is today the country of Moldova, located between Ukraine and Romania.

The settlers fled Germany because the enormous social upheavals of the Napoleonic era threatened their devout religious lives. Many were Pietists.[231, 233, 234] Some Swabians came via the Duchy of Warsaw, where they had earlier settled at the invitation of the Prussians. Whatever their origin and route, they were driven mainly by religion.[233]

Their transplanted religion thrived in Bessarabia. Unlike in Germany, religion was not controlled by the state. It operated solely on support from the community. This strong religious commitment helped the stubborn Swabians resist both Russian and later Romanian efforts to assimilate them.

Peasant Life

In the middle of the twentieth century, Immanuel and Johanna Weiss and their fellow Swabians still lived much as their ancestors in Bessarabia had five generations earlier.[235] They were self-sufficient, religious, fanatical about farming, and frugal.

About a hundred families lived in the village of Kulm.[236] The Weiss farm was of average size, about eighty acres. They grew barley as a cash crop and winter wheat, soybeans, corn, and oats for their own use. Everybody shared the village commons for grazing by their sheep, horses, and cows.

Weiss recalled how children were important for the work on the farm: "All the children—including the girls—learned to ride horses. When the corn was cultivated, a child maybe seven to ten years old would ride the horse and lead it through the corn rows—same thing for cultivating in the vineyards. When we harvested the wheat, a child would ride the leader of the four horses so the man on the mower could pay full attention to his machine."

The peasants were mostly self-sufficient. The land was steppe without trees, so instead of wood they used animal manure for fuel. "We piled up the manure all winter, and it rotted and got soft like butter. Then we laid straw down so the manure wouldn't stick to the ground and raked out a twelve-inch layer. The horses pulled a three-foot round stone to pack the manure to just four inches thick. We would use a hatchet to slice it into blocks, which we dried all summer. It had no smell, and it burned well."

For clothing, Johanna Weiss spun wool and used it for knitting and weaving. She also used goose feathers. Every eight weeks, she would take the feathers from their bottoms and half of the feathers from the front and top. They made wonderful featherbeds.

The villagers harvested flax in late summer. They laid bundles in a pond and covered them with mud for three weeks. After washing and drying, the fibers would come loose. Three different hand tools prepared the fibers that were spun and woven into sacks and bed sheets.

Depending on family size and appetites, people made anywhere from 500 to 5,000 gallons of wine every year. In the winter, the men and their wives would drink and socialize at someone's home. The women knitted sweaters, gloves, or socks while they talked.

Strictness

There was strict discipline while Immanuel Weiss grew up. He recalled, "We had no playing cards, none in the house. They were the Devil's work and not allowed in a religious family."

The family started observing Easter on the evening of *Gründonnerstag* (Maundy Thursday). From then until the evening of Good Friday, Immanuel's mother fasted. On Good Friday and on Easter Sunday, the children couldn't touch a toy, and the family went to church twice each day.

When children misbehaved, they were punished. Weiss recalled, "We had to go to the corner of the room and kneel on corn kernels. That hurt!" His father believed the saying, "the harder the rod, the better the child." Immanuel insisted, "I didn't believe that. But because our parents were so strict, we got to know at a young age what was wrong and what was allowed." Schoolteachers also used corporal punishment, saying, "If you can't listen, you must feel," meaning that they applied the rod.

The largest ethnic group in Bessarabia was the Romanians, who had been conquered first by the Ottoman Turks and then by the Russians. When Immanuel was born in 1916, Bessarabia was still part of Russia. With the defeat of the Russians in World War I, Bessarabia declared independence. It then joined Romania.[237]

Weiss attended school until he was fifteen. In his first few years, his teachers were German. After 1925, more and more instruction was in Romanian. New laws required that only Romanian be spoken in the schools and even in the German churches. Germans, Bulgarians, Russians, and Turks were pressured to assimilate.

Germany Was Different

In 1934, Weiss went to Germany to apprentice on a farm. At the head office of the training program in Hamburg, he was given 2,000 Reich marks ($500) to deliver to the farm's manager. That impressed him: unlike Romanians, the Germans trusted people's honesty.

At the German farm, he learned techniques unknown on the peasant farms in Bessarabia. He also met boys and girls from many parts of the German-speaking world, including Latvia, Switzerland, and the Transylvanian and Banat areas of Romania.

Weiss saw Adolf Hitler speak at a peasant festival. "He was a good speaker, and the Germans believed everything he said. He promised jobs and food." On taking power in 1933, Hitler had immediately embraced grand public works such as the autobahn with four lanes, where one could drive for hours at the unheard speed of fifty miles an hour. Unemployment had been reduced almost by half. Weiss recalled during his stay, "Germany was a happy nation."

Another day, all work had to stop for a special speech on the radio. This use of broadcasts was an important part of the Nazi effort to foster a national community. In 1933, the Propaganda Ministry had pressed a consortium of radio manufacturers to design and produce a People's Radio that cost less than twenty dollars. In 1933 and 1934, 1.5 million radios were sold.[145,p67]

Weiss heard Hitler tell the people, "Starting today, we are going to have an army, and nobody is going to stop us." This was the reestablishment of a mass conscription army. Immanuel recalled, "Young people were screaming and happy."

Nazi study of public opinion in Germany during World War II showed that adulation of charismatic Hitler, not the Nazi Party, its ideology, or the Nazi state, was most important in holding everything together. Evidently, Hitler had developed his charisma between 1919 and 1933. It was not a gift from birth. He was quite widely read, and he gave people the impression of being a genius. Of course, he remained unshaken in his opinions even if he were wrong.[238]

Hitler's emotional speeches communicated strength, the threat of violence, and an exalted vision of what Germany must become. However, he was vague enough, or misleading, so that the people could often project whatever meaning was important to them. If one watches film of his speeches, it is also clear how his choice of words and his manner were such potent influences on the German people. This hold on the people was later evidenced when most were dismayed at the 1944 attempt on his life. Even after the war, in 1950, 10 percent of Germans still believed he was the greatest German who ever lived.[204, 239,p271]

Also impressive for Germans after Hitler took over was the change to a more positive respect for Germany from other nations. The Allies had harassed and intimidated the republican Weimar government and had insisted that Germany accept guilt for World War I, to the humiliation of most Germans. However, according to one historian, "when Hitler began to talk, the Allies changed their tune; the bullies of the 1920s turned into the cravens of the 1930s."[240,p175]

Weiss Becomes a Romanian Soldier

After returning to Bessarabia, Weiss had to join the army reserves and take part in periodic military exercises under Romanian command. He trained with members of other ethnic groups. The

Bessarabian population was 56 percent Romanian, 25 percent Ukrainian and Russian, 7 percent Jewish, 6 percent Bulgarian, 4 percent Gagausian (Turkish Christians), and 3 percent German.[241]

In 1938, at age twenty-two, Weiss was drafted into the cavalry. His unit had about sixty Germans and an equal number of non-Germans who had no horses. By bringing a horse to share with another soldier, Weiss could serve just six months, instead of eighteen, and then go into the reserves and return to farming.

Weiss disliked the Romanian army. It was so different from the honesty that he learned in his family. Corruption was rampant. "Bribes were expected by everyone. The sergeants were always finding little ways to make us give them money. Before watch, we would grab a few hours of sleep. When we got up, the little stars on the front of our riding boots were gone, or maybe the ones from our shoulders. For roll call at 7 a.m. everything had to be in order, so we were reported to the sergeant, and it cost ten or twenty lei to get our stuff back! The Romanians were really corrupt."

After three months of basic training, Weiss was allowed to go home for two months and the harvest. During that short time, his father died and Immanuel married his fiancé of two years, Johanna. He then returned to the army for annual maneuvers.

But in that year, 1939, Germany invaded Poland, so the Romanian Army was put on full alert. "We didn't know if maybe the Poles would escape and try to come into our country. They were being squeezed from both sides. The Soviets were attacking Poland from the east and the Germans from the west."

Many Romanian troops were sent to the Dniester River, the border between Bessarabia and the Soviet Union. Immanuel Weiss's unit was assigned to cavalry headquarters in Ismail on the Danube River, 100 miles south of the Dniester. The Romanian army remained on full alert into 1940. Johanna Weiss had to use hired help to plant that year.

The Soviets Invade Bessarabia

On June 28, 1940, Immanuel Weiss awakened to learn the Soviet Union had invaded Bessarabia. The Bessarabians had been betrayed. In August 1939, Germany and the Soviet Union had signed a non-aggression pact with a secret part on how Hitler and Stalin would divide Eastern Europe. Germany would get the western half of Poland, and the Soviet Union would get the eastern half of Poland. Bessarabia was assigned to the Soviet Union.[242]

After the Germans invaded Poland on September 1, 1939, Hitler wanted Stalin to invade from the east immediately and trap the Polish army between German and Soviet troops. But Stalin cunningly waited until September 17 so he could tell Britain and France, allies of the Poles, that he was only moving troops in to help defend against the Germans. A month and a half later, the Soviet Union annexed eastern Poland occupied by its troops, fulfilling one part of the secret pact.

But there was more to come. Stalin's next move, on November 30, was to attack Finland. The Finns admitted defeat on March 11, 1940, and evacuated more than half a million people from the Finnish territories the Soviet Union annexed. The Red Army then invaded Estonia, Latvia, and Lithuania in mid-June of 1940. Stalin set in motion plans to have one third of the people in those countries sent to Siberia and replaced by Russians.

Only one part of the secret agreement remained—Bessarabia. On the evening of June 26, 1940, the Soviet Union gave Romania a twenty-four hour ultimatum to give up Bessarabia, the eastern part of Romania that bordered the Soviet Union. King Carol of Romania wanted to resist.[232,p247] However, because of the secret agreement between Hitler and Stalin, Germany refused to support any resistance, even though Germany and Romania were allies. The Romanian government surrendered and allowed the Soviet Union to absorb Bessarabia and to move its border with Romania west and south to the Prut and Danube rivers.

Becoming a Deserter

Immanuel Weiss and the other Bessarabian Germans did not want to go with their cavalry unit evacuating across the Danube River into Romania. "We wanted to stay with our own people in Bessarabia. But if we did this and deserted from the Romanian Army, we would be helping the Soviet Union. We didn't want to do that either."

The perplexed soldiers spent the day loading the squadron's saddles, horseshoes, clothes, boots, and supplies to be ferried across the Danube. That night, Weiss and two comrades decided to desert. Early the next morning, they left Ismail on foot, heading north towards their homes.

They immediately met a Romanian patrol, but Weiss and his comrades had officer's caps, tailor-made uniforms, and high boots. They were not challenged. They reached the edge of Ismail by sunrise and hid in a field of grain. It was hot, and they were thirsty and hungry. In late afternoon, they decided to start walking, and they met some peasants who gave them water. The peasants were relieved to learn that the soldiers would not steal their horses.

With remnants of the Romanian army still retreating along the roads, the trio walked at night and hid by day in fields. However, whenever they came to streams, they had to use the bridges, and road traffic was dangerously heavy.

One evening, as darkness fell, they planned to go to the nearby village of a Bulgarian comrade to rest and get food. To evade retreating Romanian units, they had to wade across one of the many huge shallow estuaries of the Black Sea. They couldn't get to the Bulgarian village until daybreak. But they received an enthusiastic welcome from their friend, Malumin Dimitri. The Bulgarian minority had a cultural affinity with the Russians, so it was no surprise when Malumin exclaimed, "We're free now!".

Many of the Bulgarian villagers wore red bands to show they supported the Communists. The three Germans reluctantly joined in their celebrations.

Meeting the Russians

Weiss convinced his Bulgarian friend to take them by horse and wagon under cover of night to the nearest of several German villages, twenty miles away.

On the road, they soon saw lights and heard the noise of tanks. Probably Russian! They got off the road and tried to hide and let the tanks pass. But a hail of bullets came, stampeding the horses and forcing Weiss and his comrades to scatter in the dark.

They regrouped next morning and thought to recover Malumin's horses and wagon despite the Russians camped nearby. Since the trio still wore their Romanian officer's uniforms, the Russian tank officer accepted their explanation that they were deserters heading home and gave them back Malumin's wagon. It was okay except for many bullet holes, but the horses were gone. The Germans had to continue on foot. Weiss and his comrades had been lucky to escape with their lives. Some overtaken Romanian units were shot at and sometimes captured and imprisoned.[243,p234]

After a few miles, they met Russian cavalry. Rather than hide, Weiss decided he'd best act friendly towards them. Weiss knew some Russian.

"Drastite tovarichi." (Greetings, comrades.)
"Where are you going?"
"I want to go home."
"Yes, go on home."

That evening, they arrived at the last of the Bulgarian villages to stay with another comrade from their cavalry unit. The next morning, Soviet infantry marched into the village, and children greeted them with flowers. Weiss recalled, "It made me feel so bad."

Weiss had often heard news coming across the border from the Ukraine: how the Soviets sent men and grown boys to penal camps in Siberia, never to be seen again, or sometimes arresting a whole family; how people were reduced to eating soup made from cowhide or sheepskin and weeds; how some peasants even had to eat the bodies of the dead to avoid starving.[244]

Weiss was filled with fear and foreboding. He left the Bulgarian village by himself and reached the first of the German villages the next day. He then phoned his brother and got a ride home to Kulm.

Under Communism

Weiss was happy to be with his wife Johanna and their eighteen-month-old daughter, but he was worried. "We knew that the Communist system didn't permit free peasants. They forced them into collective farms. What would happen to us?"

Earlier, in 1917, the Soviet state had abolished all private landholdings. An estimated five million had died in the terror when the Soviets first took over. From 1930 to 1937, Stalin forced peasants onto collective farms, with eleven million deaths from famine and violence and an estimated three and a half million deported to penal camps." The prospects for independent peasants like Immanuel and Johanna were not good.[234, 244]

Weiss found the store shelves in Kulm empty: no clothes, no pepper, no matches, and most important, no salt. Lacking refrigerators or freezers, people preserved their meat by layering it with salt in a stone crock or wood tub. They also needed pepper for making smoked ham, wurst, and bacon. It was summer, not a good time to butcher animals. The peasants had cattle, pigs, and sheep but couldn't slaughter them because the meat would quickly spoil. Weiss wondered, "Maybe the Russians arranged the shortage of salt and pepper for that reason."

The German men were ordered to report to a neighboring village to build an airfield, but that was only for two days. "Maybe they didn't want us to learn too much about what was being built."

The wheat, barley, and oats were ready for harvest. But the order came that the state would take almost all, leaving too little for the animals and people to survive until the next harvest. Weiss believed the Soviets wanted the peasants to give their animals to the state. When they had nothing left, they would probably be forced to become employees of a state farm—or sent to Siberia.Rescued by

Hitler

When rumors circulated that Germany would rescue all the Germans in Bessarabia, the villagers in Kulm updated their *Stammbaum* or family tree to prove their ancestry. The good records kept in the local churches helped.

Hitler's offer to save the 90,000 Bessarabian Germans was part of a larger plan for the *Volksdeutsch* or ethnic Germans, first announced in his speech of October 6, 1939. Hitler needed German settlers to bolster the minority German population in the part of Poland he had annexed.

Apparently, the Soviet authorities hoped that not more than 30 to 50 percent of Bessarabian Germans would opt for transfer to Germany. The Soviets tried to prevent the younger people from registering for the repatriation lists. Nearly a third of the Bessarabian Germans were children under 14.[245,p200] Soviet representatives demanded the children appear in person before the German-Soviet commission. A German observer noted that the Soviets "attempted by continuous questioning of the children to separate them from their parents, and frightened them so." The hope was that they would stay. The effort was a failure. [245,p18]

Weiss recalled, "About the fifteenth of September, four men arrived, a chauffeur and three SS officers. Everybody was ecstatic. We were saved! Everyone in Kulm registered for resettlement, even the two families where there was marriage between German women and Bulgarian men." That and some creative genealogy might explain why more Germans were repatriated than their numbers earlier recorded in the census. Germany's leniency also allowed perhaps ten thousand White Russian emigres, Ukrainian nationalists, and others who were anti-Soviet to be evacuated as Germans.

Land, houses, farm buildings, animals, machinery, and crops were evaluated. "We had to leave everything, but the Soviets were going to pay for what we left behind."

After three months of living under Communism, Weiss and his family were eager to leave. On October 2, 1940, forty buses came to

Kulm to take women, children, and older men to Galati, a town on the Danube in Romania. Able-bodied men left four days later, driving one wagon per family and taking the allowable fifty kilos (110 pounds) of baggage. From Galati, the Bessarabian Germans sailed up the Danube River to Prahova where they boarded a train to refugee camps in Germany.

Resettled and Drafted

About 2,000 Germans from Kulm and a neighboring village were housed, thirty to a big room, in a depressing refugee camp near Schweinfurt. Everyone over fifteen had their blood type tattooed under their left arm. Weiss and his family members were sworn in as naturalized German citizens and given papers to receive a farm in the eastern part of Germany, Warthegau.

Germany had conquered half of Poland and half of that was then annexed into Germany to become the province of Warthegau. Poles and Jews were forcibly removed from urban centers. German refugees from the Baltic states that the Soviets had occupied were resettled there. The small farms of the Polish peasants were combined into larger holdings and their owners deported to the part of Poland not annexed by Germany. Only a few families were left in peasant villages to become laborers for the German farmers.

Most of the new German farms were administered by trustees. The German farmers were not fully independent. Germany paid for repairs and upgrades of houses, stables, and barns to meet German standards. The farms were also stocked with implements, grains, and livestock. The area was to become important in providing foodstuffs for Germany.[245]

In February 1940, the Weiss family was assigned to a ninety-acre farm in the hamlet of Klein Lohe, south of Posen in the province of Warthegau. Their farm was about the same size as the one they had left in Bessarabia.

In April 1940, Germany invaded Denmark and Norway, then Holland and Belgium in May, and France in June. Hitler only had

Great Britain to defeat. However, the British refused to accept German proposals for an agreement that would divide Europe into mutual spheres of power. The swift victories until this time made Hitler immensely popular among most Germans, including the few who had earlier feared his aggressive attitudes. However, the British refused to buckle despite more and more German bombing raids.

There was some good news for Germany as Hungary, Romania, and Slovakia joined the Tripartite Pact of Germany, Japan, and Italy in November. However, Italian efforts to conquer Greece and Egypt's Suez Canal were blunted, and German troops had to help. After Bulgaria joined the Tripartite Pact, Hitler moved to take over Yugoslavia. This was an impulsive decision that would hinder Hitler's next move, against the Soviet Union. Some Germans started to fear the nation would one day collapse under the weight of its excessive expansion.[238]

Hitler ended the non-aggression pact and attacked the Soviet Union in June 1941. He thought German power could easily defeat the Russians. There were three goals: 1) to get *Lebensraum* or more territory for the German people, 2) to rob England of its last continental support, and 3) to wipe out the "Jewish-Bolshevik" enemy including the Soviet elite and all Jews there.

Germany launched a *Blitzkrieg* invasion aimed at destroying most of the Red Army and capturing Moscow before winter. However, German casualties were high. Further, instead of the 200 Soviet divisions Germany expected to face, there were 360 divisions. They put up stiff resistance despite over 460,000 killed and 3.3 million taken prisoner by the end of 1941. During this time, German SS and army units murdered 500,000 Jewish civilians in the conquered Soviet territories.[145,pp184,185, 238,p120]

German public opinion was not greatly impressed by the initial victories in Russia. A Nazi report for Stuttgart by the *Sicherheitsdienst* (intelligence office) revealed that, "Everywhere one finds the opinion that 'in this war *the little people are the losers once more*.' They must work, do without and keep their mouths shut while others can afford anything with money, connections, license, and ruthlessness. *'Everything is just like it was before... Can one still speak of a people's community?'*[238,p122]

Because the German army was not prepared for winter fighting, the Red Army stopped the Germans short of Moscow. In 1942, German troops advanced hundreds of miles, mostly in the southern part of the front where the Russians were weakest. The Red Army had five million casualties in the first fifteen months of the war but held on and staged a deadly counter-offensive during the winter of 1942-43 at Stalingrad.

The Red Army lost nearly 500,000 men in the seventy-two days of fighting at Stalingrad, while Germany lost nearly 300,000, a loss that Germany could ill afford. The debacle at Stalingrad was followed by the German loss of Tunisia and 250,000 men taken prisoner in North Africa.

After Stalingrad, a rare public German protest took place, the White Rose action. Swabians Hans and Sophie Scholl from Ulm were caught spreading leaflets at the university in Munich. They were executed by guillotine immediately after sentencing.

The subsequent July 1944 unsuccessful assassination attempt on Hitler was a similar misreading of expected public support to overthrow the regime. While people were disgusted with the Nazis, they still adored Hitler. Nazi intelligence reports showed that even critics of the regime rejected this attempt at murder despite the war being clearly lost at that time.[238]

The debacle at Stalingrad meant that Germany desperately needed more troops, so the draft expanded to snare even essential farmworkers like Immanuel Weiss. He was drafted into the German Army on March 25, 1943, along with many other ethnic Germans from Russia, Romania, Hungary, and Yugoslavia.

Weiss was twenty-six, but many others were up to ten years older. After two weeks of training in East Prussia, Weiss thought, "If it's all like this, I can stand it." He was assigned to the light cavalry, which used bicycles and motorcycles. He was also given some infantry training and then sent near Paris for six more months of training. "They said I could get my Romanian corporal's rank back, but I wasn't crazy about that. The higher up you are, the harder you have to work. I didn't want promotions. I just wanted to come back home alive."

While Weiss was training during the summer of 1943, the Red Army started pushing the Germans back. By fall, there were only 2.5 million German troops left in the east to fight the 5.5 million in the Red Army.[246,p62]

The Partisans

In September 1943, Weiss was sent to protect German trains from Soviet partisans. These were Army stragglers overrun in 1941 plus some local peasants in the forests of Belorussia, the northern part of the front. At night, Soviet planes would fly into secluded airstrips with weapons, munitions, and supplies, plus demolition experts and trainers. By mid-1943, more than 200,000 partisans were daily targeting German road and rail traffic with hundreds of attacks.[246,p86, 247]

Weiss had to guard a rail line near Karamischewo, a few miles east of Pskov. Russian partisans came out of the forests at night to plant land mines to derail the trains. Weiss and the other German troops were protected with bunkers placed along the rail line about a mile and a half apart. Starting at dusk and every half hour, two men left a bunker and went east, and two men went west. They would pass another patrol halfway, go on to the next bunker, sign in, and then return. They alternated two hours on duty and two hours off all night.

When the Germans on patrol spotted a mine, they marked it but just kept walking. They did not want to give their discovery away to the partisans hidden at the end of the wires connected to the mines and waiting for a locomotive. Later, a special armored train would remove the mines. One night, seventy mines were found in Weiss's small area. Sometimes the partisans attacked. At Christmas, a new unit from Germany was surprised inside their bunkers and killed.

On January 14, 1944, the Red Army moved to break the terrible two-year siege of Leningrad. They had a three-to-one advantage in troops and a six-to-one advantage in tanks, self-propelled artillery, and planes.[246,p97] In two weeks, 14,000 Germans were killed and 35,000 wounded. By the end of the month, the German 18th Army was close to collapse.[248,p251]

A member of the elite Waffen-SS (Army SS) troops claimed of the situation on the East Front: "We weren't outfought, but we were outnumbered, overwhelmed, pushed to the wall by sheer weight." This would be a continuing refrain.[249,p172]

Fighting at the Front

Five percent of all troops in the rear were hastily gathered to reinforce the front line. Then Soviet troops from the south suddenly attacked near Lake Ilmen. Weiss and his comrades were sent there to form a new unit, the 368th Grenadier (Infantry) Regiment. They arrived at night to replace casualties on the main battle line along a railroad.

The area was swampy with a high water table so bunkers had to be on the rail embankment, and only rifles and machine guns, not heavy weapons, could be used. The Germans got supplies of food and ammunition by armored train, while the Russians were stuck in the swamp and forest.

Each night, some of the German soldiers took turns at an observation post on the far side of the tracks. Even with the light reflected from the snow, they could not see into the forest. On the third night he was there, Weiss and the others had to repel a Russian attack.

The next day, Weiss and four others were ordered into no-man's land across the railroad to collect guns and ammunition from the Russian dead. The Russians shot at them while covering fire was returned. The next day the Russians brought up a big supply of munitions, and Weiss had to join a team that crept out at night to detonate their stores.

Retreating

A few days later, the Germans had to pull back close to the city of Dno. Weiss and his comrades dug in around the main road. Early in the afternoon, Russian tanks suddenly appeared about a third of a mile away. They rolled across the fields, six to the left and seven to the right of Weiss's unit. With only sixty men and four tanks, the Germans were badly outnumbered. They called for air support, and Stuka planes came and destroyed every Russian tank.

Early the next morning, Russian soldiers filled the countryside, and tanks carrying more infantry rolled in. Weiss and his comrades were outnumbered five to one. The Germans held their fire until the Russian tanks were fifty yards away. They blew up the tanks one by one, and the rest fled. Weiss and his comrades kept shooting as waves of Russian infantry came over a rise. They killed hundreds, their brown uniforms covering the snow where they fell. After nearly four hours, the Russians finally stopped coming.

Toward evening, a sergeant ordered the troops to hold their position a couple more hours. But could that be right? The sergeant was Lithuanian and didn't know German well. It was getting dark. Finally, the unit's messenger came and yelled, "Get out! The others all left already." Fortunately for Weiss, Hitler now no longer ordered positions held at all costs as he earlier had at Stalingrad.

The Germans ran out of their foxholes. They heard the voices of Russian commandos followed by the explosions of incoming hand grenades. The soldier next to Weiss had his leg blown off. They dragged him back in a snow-boat. One man from each group was ordered to stay to provide covering fire for the retreat. These men were never seen again.

Weiss recalled, "Earlier, it was me that had to stay behind while the others retreated. You had no choice. You had to obey orders. If not, you'd be court-martialed." Soon German reinforcements came on snowshoes to stop the Reds. Weiss and his unit retreated through several cities until they reached a safer position where they stayed for several weeks.

Wounded

Weiss's unit was sent to the front again at the end of March 1944. They found themselves facing Russians a few hundred yards away across a small valley. The German foxholes were hidden behind bushes. Weiss was lifting several trays of food and bottles of coffee from their mess wagon when a mortar shell hit 100 yards away, then a second, closer, and a third, right next to the wagon. Weiss was hit with shrapnel in his legs, and he crawled through the foot-deep snow on his elbows until he got first aid. He was taken that night to a hospital where all but nine small pieces of shrapnel were removed.

It was nearly two months until Weiss could walk again. Two weeks' recovery leave was to begin June 10, but four days before then, the Allies landed in Normandy. All leaves were cancelled.

Weiss traded his large collection of saved whiskey rations to a sergeant in order to get assigned helping in a hospital in Riga, Latvia. However, as the Red Army came closer at the beginning of July, the hospital closed. Because his wounds had opened up again from the work, Weiss was sent back to Germany.

At the hospital in East Prussia, officers ordered him to the front. They relented when Weiss told them that it was his birthday. Instead, he was sent to Thüringen, Germany, to recover. There he convinced the commander to give him two weeks' recuperation leave plus two weeks' work leave because he had not been on his farm for over a year. On the farm, his wounds broke open again. When he returned to his unit, he found that many of his comrades had been lost fighting the Russians. Because of his wounds, Weiss was assigned to a motorized antitank unit. That was okay with him. He figured every day in training made the war shorter. He prayed a lot, too. He wanted to survive.

Starting in October 1944, Weiss spent five weeks in Kolberg, Pomerania, for training on antitank guns. He got a short leave to go home, and when he returned to duty, he found the rest of his unit had gone to the Eastern Front. No one had been able to reach him on the farm. He spent Christmas 1944 quietly in camp.

During the last half of 1944, while Weiss was in training, the Red Army by-passed most of the German troops in the Baltic states and made a strong push southeast, heading towards Warsaw and East Prussia. German strength was ebbing quickly. Between June and November 1944 the Germans lost 1,457,000 men.[248,p412] Hitler called up the *Volkssturm* or home guard consisting of boys as young as sixteen and men as old as sixty.

When the Red Army first reached East Prussia in October, they put up posters telling their troops: "You are now on German soil; the hour of revenge has struck."[246,p188] A leaflet ordered the troops: "Kill. Nothing in Germany is guiltless, neither the living nor the yet unborn."[250,p65] Captain Aleksandr Solzhenitsyn noted "all of us knew very well that if the girls were German they could be raped and then shot."[251,21] He added, "Whoever still a virgin, soon to become a woman; the women soon to become corpses. Eyes bloody, already glazed over, pleading: 'Kill me, soldier.'"[206,p193]

The Soviet Union and other Slavic countries, including Poland, had lost thirty-five million people. The Germans had been brutal. Thus, retribution towards German civilians would be terrible.

After a pause, the Red Army resumed its offensive in January 1945. They outnumbered the Germans more than two to one overall. The Russians easily broke through German lines and headed towards the province of Warthegau.[248,p417]

The War Comes to Johanna Weiss

After her husband had been drafted in March 1943, Johanna was mostly alone with the Polish workers who helped on the farm. There were three other German-Bessarabian women in the village, but the rest of the people were Polish. In 1944, ethnic German refugees came to the village: three families from the Ukraine and a woman and two boys from the Crimea. They were ignorant of the horrors experienced by German civilians in East Prussia. They were isolated on the farm and busy working to meet the increasingly stricter production quotas set by the German government.[252, 253] There were almost continuous air attacks by the end of 1944. Soviet fighter planes were strafing the adults and children working in the fields.

On January 20, 1945, just after Johanna had bathed her three girls and put them to bed, someone knocked at her window: "The Russians are close. You must be in Karlshausen in two hours." The January offensive by the Red Army had been massive and swift, advancing at a pace of 20 to 30 miles a day.[248,p424]

Johanna had made preparations to flee by weaving a flax cover to put on a frame atop her wagon. The Polish hired hands helped her put up the covering and gather flour and grain. She took a five-gallon can of milk, some lard, cooked sausages and ham, and all her bread. She gathered the most necessary clothing, and two feather beds, one to put underneath and another on top of the three girls. Pots and pans and furniture and everything else had to stay behind. Most of the wagon was loaded with sacks of oats for the horses and rye flour for bread.

One of the younger hired men asked her to come to the attic and help hold sacks while he filled them with rye flour. Fortunately she didn't. The other hired men told her the man had intended to murder her.

The wives of the hired men, however, were all helpful. "They used to get food from us, and a boy and a girl stayed with us and ate at our table. We treated them as equals and they knew that. The German government said we shouldn't, but we felt if we worked together, we would eat together. That's why these people helped us get ready."

Millions of other Germans in the eastern part of Germany also feared the oncoming Red Army, but some probably found it very difficult to leave their ancestral homes. Silesia, East Brandenburg, Pomerania, and East Prussia had been almost entirely German for five, six, or more centuries. Some people probably thought they could endure Soviet occupation. They did not know that Poland and Russia would annex these territories and take away their homes and farms.

The ethnic Germans living in minority enclaves in

Czechoslovakia, Hungary, and Romania would also be forced to give up homes where generations of ancestors had lived. However, some of them had behaved arrogantly toward their non-German neighbors, even criminally, when the Nazis had occupied their countries. Some fled, but others probably felt they were innocent and could stay.

Johanna Weiss's hands were clean; she had gotten along with her Polish neighbors. It was easy for her to decide to flee because she knew what it would be like to live under the Communists.

Nearly five million Germans in the eastern part of Germany wisely left their homes early in 1945 in the dead of winter.[254,p326] In Karlshausen, the farmers—women, children, and elderly men—gathered to begin the trek westward, traveling almost continuously on snowy and icy roads. It was the coldest January in years.

Johanna Weiss recalled, "On the wagon in front of us there was a woman with many children. She would walk to keep warm. When she couldn't walk any more, she would sit on the wagon. No one knows if she fell asleep or what, but she dropped to the road. Dead. Her children cried so. But her body had to be left. Everyone had to keep going. Many people were left dead on the side of the road like that, some bodies covered with a cloth, some just lying there." People jettisoned belongings as well as dead babies into the ditches.[206,p194]

After almost two weeks, they arrived in the village of Züllichendorf, thirty miles south of Berlin, where they got a room. "We lived with a family, a man and woman, their daughter, and their granddaughter. People didn't want to take in refugees, but they were forced by the German government. That's why they didn't always have warm feelings towards us. And we paid no rent. We lived for free because we lost everything from the war while they were still able to keep something."

Under the Bombs

Johanna learned that Immanuel's sister, Elise, had also found refuge in the area, so she set out to get her. She took the children along for a three-day, twenty-five-mile trip by horse and wagon. As they neared Berlin, they heard air raid sirens howling. "We hurried into the woods, taking the horses with us. German troops came and told us to lie down for safety. Then we saw the planes, flying low in great masses. We prayed they shouldn't drop their bombs on us. I'll never forget how the earth shook under us. We thought the end of the world had come. Our horses tore their harnesses from fear.

"While we were lying there, I noticed how the ground and grass smelled so fresh. Nature was beginning to come alive again; spring was coming! I thought how nice spring would be if only we had peace. Here nature was coming to life, and only a few miles away, death. People were being burned by the phosphorous bombs or buried in the shattered homes. How can we ever understand the hatred of people, and the leaders of people, the governments? They were worse than wild animals that, after all, kill only when they are hungry." Johanna was overjoyed to collect Immanuel's sister and get away from Berlin.

By this time, Berlin was bombed almost daily by American and British planes, and, on February 13-14, just seventy-five miles south of Johanna's location, one of the most devastating air raids of the war took place at undefended Dresden. The city was packed with refugees fleeing the Red Army, and the raid killed thousands of people.

In the middle of April, unexpectedly and without a fight, Johanna heard about an army speaking a strange language that appeared in nearby Luckenwalde. The Americans stayed only two weeks, and then the Russians came.

Johanna and the Russian Occupation

Johanna Weiss recalled, "The first thing the Russians did was to look for watches, then they raped the women. They came looking for the daughter of the lady where we lived. Well, I was pregnant with our fourth child, to be born in a little more than a month. The daughter was lucky, too. She lived with the hired man, a Serbian, and he hid her under the straw by the cows."

"People tried to stay inside as much as possible, to keep away from the Russians. Many others fled. They disappeared overnight and left everything behind."

Many women were not as fortunate as Johanna Weiss and her little girls. The Germans have collected volumes of depositions, many fully corroborated, of what happened to civilians at the end of the war, stories of women forced into brothels or raped regularly for months by strangers breaking into their shelters.

Some German women did survive because of acts of kindness. One deposition tells how a farmer's wife watched Russian troops pull her husband from their wagon and shoot him, but the Polish women in the wagon, the family's farmhands, hid her with their shawls.[255]

The general feeling at the time, however, was that the Germans were collectively guilty of all the brutalities committed by the Nazis. President Roosevelt had written Secretary of War Stimson in 1944: "The German people as a whole must have it driven home to them that the whole nation has engaged in a lawless conspiracy against the decencies of modern civilization."[250,p14]

The Czechs, after suffering a brutal Nazi occupation, got revenge by taking over the Nazi concentration camp at Theresienstadt and replacing the Jews with ethnic Germans. One of the former Jewish prisoners reported: "Many among them [the ethnic Germans] had undoubtedly become guilty during the years of occupation, but in the majority they were children and juveniles, who had only been locked up because they were Germans...The people were abominably fed and mistreated, and they were no better off than one was used to from German concentration camps."[256,p76.]

Sixteen million Germans lived in the part of Germany subsequently annexed by Poland and Russia, and more Germans lived as minorities in other Eastern European countries. About two million of them were killed, either near the war's end or in the months afterwards as the people of Eastern Europe exacted revenge.

Immanuel Fights on the Western Front

During this time, Immanuel was also trying to stay alive. On January 2, 1945, he was assigned to the Western Front. "I knew I had to go, and I was happy. Someday soon the war would be over, and on the Western Front there was some chance to live. If I survived the fighting and was captured, I'd be dealing with Christians."

Weiss was sent to Huchelhofen on Germany's western border. This was near Aachen, which was the first German city to fall to the Allies on the 21st of October after a 39-day battle.[206] Weiss and his comrades faced the British. The German soldiers slept in the cellars of badly damaged houses in the town. The German antitank guns could not penetrate the armor of the British tanks, so Weiss was outfitted with a rocket launcher which would be effective if he got very close to the tanks.

Everyone knew it was no use fighting. "If we were attacked, we would fight. If not attacked, we'd slowly move back." This had to be done carefully because in the last months of the war many soldiers were shot for refusing to fight.

After one firefight, a white flag went up. "We went out to get our wounded and the British got theirs. We never saw this before, not in Russia. When you were wounded there, you either got immediate help from your buddies, or it was curtains."

Weiss kept retreating. He became depressed with no word from his family. He knew the Russians had overrun Warthegau in the East. "Our families were lost. Was life still worth living?" Yet, no one talked about surrender because they did not trust the strangers in their unit not to report them. They would be shot for such talk.

Surrender

On Easter Monday (April 9th), Weiss and a few men were enjoying dinner with some local people when the word came, "the British are nearby. We must give up the village." Weiss's unit moved into the woods and dug foxholes. The next day a nearby farmer who had been quartering their lieutenant told them that he had left. As they walked away from the farm, they came under fire. Weiss immediately waved a white handkerchief.

They were taken to a sixty-acre field full of prisoners (POWs). "It was raining, but we just brushed the water and mud away, laid one coat on the ground, and used another coat as a cover. We were two men together, lying on our sides, first one side and then the other, to keep a little warm." For five days, their daily ration was ten crackers. Then they went by train through Holland to Belgium. Along the way, people threw rocks into their open rail car and poured boiling water down from bridges.

When they got off at Waterloo in Belgium and marched through town, people threw rocks and water on them from the upper stories of buildings. To their left, the Belgian guards hit them with bayonets. A weakened prisoner fell out of line and was shot, but on their right, the British protected them.

Weiss was held at the POW camp in Waterloo with 60,000 men. One day, British officers lined up everyone and ordered shirts off They looked under arms for the telltale SS tattoos. Weiss was reassured that his blood-type tattoo was no longer visible.

In June, the British released nearly half a million POWs to help bring in the German harvest.[257,p132] Each got sixty Reich marks ($15) for their work except for those who had lived east of the new German border on the Oder River. They got nothing. Weiss said: "I guess since Germany lost all that territory, the people like us who used to live there weren't considered Germans anymore." Weiss and other troops from the lands that Germany lost—Prussia, Pomerania, Silesia, and Warthegau—were seen as refugees, not German soldiers.

The British had released Weiss after two months. The Soviets held some of their POWs for five years in Siberia, and 1.5 million of those 3.5 million prisoners died. Many more Soviet POWs had died at the hands of the Germans. But on the Western Front, the Allies treated their POWs legally.[258]

Weiss had now lost two farms, one in Bessarabia and one in Warthegau. Such was his commitment to farming that he would spend a couple decades, including a move to a new continent, in order to become an independent farmer once more. His commitment, and that of his wife, Johanna, made them persevere so they could once more resume the life of an independent farm family.

Chaos In Germany

In defeated Germany, there was chaos. Some fifteen million had fled the Allied bombing of the cities for rural villages. They returned hoping to learn if family and friends were alive and if their homes still existed. The remaining Germans in the cities traveled to the countryside seeking food. Two million fled their homes in the Russian zone to cross into the western zones of occupation, while a million went in the opposite direction. A British correspondent noted at the end of August 1945 that "there are eight million homeless nomads wandering about the areas of the provinces near Berlin."[254,p327]

Weiss started to search for his family. He went into the nearby big cities, Hildesheim and Hanover, and checked the Red Cross lists of refugees. In September, Weiss traveled to Lüneburg near Hamburg, riding on top of a coal car because he had no money for a train ticket. At the refugee center run by the Red Cross, he put his name on a makeshift bulletin board on a fence.

Since May, there had been no regular mail service to the Russian Zone of Occupation. So Weiss asked a soldier traveling there to take a letter addressed to a Berlin woman who had been evacuated to the Weiss farm in Warthegau during the war. The letter eventually got to her, and she knew where Johanna and the children were. When mail between the occupation zones resumed in late October, Weiss learned that his wife, the three girls, and a baby boy born four months earlier were alive and safe.

Most of Immanuel Weiss's close relatives survived the war. Two of his brothers were too old to fight, but the other one was forced to join the Volkssturm at age forty-seven. When the Russians captured him, they took him to a barn where they shot a dozen prisoners, but he was among twenty or thirty who were spared.

The husbands of Immanuel's three sisters were not so lucky. One was killed on the Eastern Front, one starved to death in a Russian prison camp, and one came back an invalid. Johanna Weiss lost her only brother in the war. Of the 351 men and youths from Immanuel and Johanna's Bessarabian village of Kulm who took part in the war, only 174 survived.[236,p204]

Rescuing the Family

Immanuel Weiss knew his family was alive, but getting them out of the Russian Zone of Occupation would be dangerous. The Russians were imprisoning or shooting Russian-born men like Weiss who had fought against them.[206, 257]

At the border of the Russian Zone of Occupation, thousands waited to cross in each direction. At 10 a.m. the Russians opened the gate to those coming into their zone, and at noon the Russians would let out the same number.

Weiss and a couple of other Bessarabian men met up and crossed into the Russian Zone. He eventually reached Züllichendorf. "What a joy to see each other again," Weiss recalled. By God's grace he had found his wife, children, and three sisters safe.

They immediately made plans to travel to the West. They could not be sure the border would be open a single day more. They packed their few possessions and went to the railroad station to rent an empty boxcar for the ride to the border, but it didn't work out. The Russians sent them back.

Weiss had already gotten a British Zone of Occupation residence permit. So he and his brother-in-law went into the British Sector of Berlin and signed up for a transport across the Soviet Zone into West Germany. The British occupation authorities told him he might be able to make the trip in a month or two. In the next weeks, Weiss made several trips back to Berlin to see if he had gotten a spot. He had to do this carefully to keep village authorities from discovering he was planning to escape the Soviet Zone with his family.

Finally, Weiss learned that a British transport would leave on January 3, 1946. They had a week to get ready. They hired a driver and an old army truck. They also took Immanuel's sister with her two children and Johanna's sister-in-law with her three children, even though none of them had residence permits. All thirteen hid in the back of the truck, covered by a canvas tarp, for the trip from the Soviet Sector of Berlin into the British Sector.

Weiss's residence permit only covered six—him, his wife, and their four children. When the clerk for the British transport asked for his permit, Weiss showed it but pulled it back quickly. "How many people?" she asked. "Thirteen," he replied.

The same thing happened after they crossed the Soviet Zone and were about to enter the British Zone of Occupation in West Germany. No one saw the residence permit was for only six people.

But as they prepared to cross the border, Russian soldiers came down the line of boxcars and, one by one, robbed the refugees of their luggage. They called out, "*Otkreu dwer*," open the door. Weiss stalled and asked who they were. "Open or we'll break it down."

When a Russian soldier entered and asked where they were going, Immanuel replied: "Home." Just then, the train started to move, and the Russian jumped out. Weiss recalled: "The Russians just ran out of time."

Time Zero

The Germans refer to the period after the war's end as "Time Zero."[257, 259-261] Devastation and economic collapse were worse than at any other time in German history. Eventually, ten million refugees—mostly women, children, and elderly men—crowded into the occupation zones of the Western Allies. The refugees came from the Eastern part of Germany annexed by Poland and the Soviet Union and from the ethnic German parts of Poland, Czechoslovakia, Hungary, Romania, Yugoslavia, and Russia. In addition, there were untold numbers of Poles, Balts, Ukrainians, and others who refused to return home to live under Communism.[257,p191, 262]

Shelter was scarce. Most West German cities were more than half destroyed. Many residents were lucky to live in cellars underneath ruined buildings. Twenty million were homeless. The Weiss family ended up in the northwestern, industrialized part of Germany occupied by the British, where the average dwelling space for an entire family was nine by six feet.

Everything was scarce. Germany's coal mines could produce only a quarter of their earlier output for winter heating. There were few German men available to work. The 1946 German census counted 36 million females and only 28 million males.[263,p66]

Food was also scarce. The primary agricultural regions of Germany were not in West Germany but in the East. In the British Zone, daily food rations were set at 1,550 calories a person, about half the normal consumption in a Western European country.[259,p52] In practice, the basic ration in the British zone was set at 1,048 in the year after the war.[257,p133]

The food crisis worsened. The Soviets had agreed at the Potsdam meetings of August 1945 that they would ship food from the agricultural territories they occupied in exchange for industrial equipment and products from the British and American Zones. This was in addition to the industrial equipment the Soviets were already stripping from their own zone as reparations. However, they reneged on the deal and didn't ship the food, so the British soon had to cut the rations for Germans to 1,015 calories a day.

Millions of Germans expelled from the East continued to flood the Western zones of occupation. Two years after the end of the war, people in West Germany still averaged only 1,040 calories a day (while American soldiers averaged 4,200).[257,p145]

Starting Anew in Swabia

After reuniting his family, Weiss decided they would return to Swabia, where their ancestors had lived before migrating to Bessarabia 150 years earlier. He recalled, "The people of Prevorst were just small farmers. The largest farm was only twelve acres. They had to work hard to survive, but almost without exception, they were good to us. From this person came a bench or a chair and from that person a table or a bed. It was soon livable. Gradually we became known to the people, and we got work."

The second day there, Weiss chopped firewood in exchange for food. He also started working at the inn for $7.50 a week. On Sundays, there was an additional payment of a large loaf of bread and a couple pounds of meat or canned wurst.

It was hard to get additional food. Government ration stamps permitted the purchase of 2.2 pounds of bread, seven ounces of meat, and five ounces of butter a week per person. "It was a lot of bread, but not much to put on it. Sometimes I might eat a whole pound of bread at a sitting, bread with nothing on it. Other meals were just plain boiled potatoes. You had to get extra food some other way. That's where the ration of three cigarettes per day helped. We would trade for food or sell them for as much as RM 5 ($1.25) per cigarette."

Sometimes Immanuel, Johanna, and the older girls, aged eight and five, helped one of the farmers with haying. They were paid with a meal or a big loaf of bread. Johanna and the children also gleaned the fields after the harvest, picking up twenty-five pounds of grain from the ground. At home, they would pound the heads of grain, clean out the chaff, and then take it to the mill to be ground into flour. They also helped to harvest potatoes and were paid in kind.

Besides the farm work, Weiss worked winters as a logger for the state where he got wood for cooking and heating. He also trapped voles and moles, burrowing animals that chewed the roots of fruit trees. He made more money at that state job than any other job because it was piecework. In 1947, he also started breeding domestic mice for a company as a sideline. "I was always looking for some kind of work. At the time, there was no welfare available. All of Germany was in need of welfare! This made us, and Germany, tough. People used to say that even if you were Jesus Christ, it didn't matter. You still had to help yourself."

Slowly, Weiss and his family moved ahead. They bought ten baby chicks and raised them to produce eight or nine eggs a day. Next, they bought a young pig. To get feed for the animals, Weiss worked for a farmer and took home his kitchen garbage. The children also cut weeds along the road.

They saved to buy an old bike so Weiss could ride to his jobs. Next, they bought an old bedroom set. Slowly, they got more things, like cooking pots. In 1947, they bought a used radio. But Immanuel and Johanna and the five children still lived in two rooms and they shared the kitchen with two other families.

The currency reform of June 20, 1948, was a major blow. Every German could change only sixty of the old Reichsmarks into sixty of the new Deutschmarks, and the rest of the family's 13,900 Reichsmarks ($3,475), which they had scrimped to save over the years, became worthless. Years later, they would get back about ten percent of that money, but most of their life savings was lost.

The currency reform improved the German economy. It did away with most rationing and eliminated the black market. In 1948, Weiss started work in a new factory making cement blocks and roof tiles. The next year, Weiss got a better job, as a laborer on a road gang. The work was exhausting, and the commute meant he was away from home for thirteen hours a day. Most refugees like Weiss had a harder time recovering from the war than did native Germans.[264]

Leaving Germany

Weiss recalled, "I couldn't really feel good in Germany. I wanted to have a place I could call my own. I wanted to have some land where I could be happy and plant things again. I'd go to any country to do it. Even to Australia or America."

Weiss told me that his brother later made a study of what happened to the thousands of Bessarabians who settled in Germany. "Only a single peson got to be an independent farmer. You have to wonder, is that a good thing?"

Immigration was not easy for the Bessarabian Germans.[254, 265] America still viewed them as German enemies. At first, the only

Germans allowed into the U.S. were German Jews who had survived the Holocaust or American-born children of Germans who had been trapped in Germany by war.

However, the Cold War with the Soviet Union changed things. The Displaced Persons Acts of 1948 and 1950 opened the U.S. to people who could not return to their homes in Communist-occupied countries such as Latvia, Lithuania, and Estonia. Farmers were especially welcomed. Eventually, even Germans found they could again get visas under the German immigrant quota.

Sponsorship was more difficult. Someone had to guarantee housing and employment for the immigrants, and the Weiss family now included five children. In the summer of 1951, Weiss heard the Church World Service was sponsoring immigrants to America. He applied, and after a year the family's name came up. The Weiss family was fortunate, for there were still nearly ten million other German refugees in West Germany in mid-1951.[254,p330]

Making use of special immigration programs involved a tedious and extensive bureaucracy. Every potential immigrant was scrutinized and a report written.[266] Weiss recalled his interrogation: "First, they asked if I would fight for the U.S., and I said I would. Then they told me that I shouldn't plan on getting any help from the government, and I told them that I was coming to work. Finally, they asked if I wanted to change my name. Some Germans changed their names for political reasons, but not me." On April 17, 1952, Weiss and his family left on an American troopship for "the land of unlimited possibilities."

Settling in Iowa, their biggest immediate problem was finding housing. They were able to move in with the Lutheran pastor, who had a large house. Johanna cooked for both families and baby-sat for the pastor's children. Immanuel and Johanna paid the utilities.

However, Weiss recalled, "We were always hungry. I worked at a plasterboard factory where there was much heavy lifting and carrying. I needed plenty of food for energy. The pastor's wife told Johanna what food to cook. For us eleven people, there was just nothing on the table. The pastor didn't work hard, and he slept in the morning, so he didn't need a lot of energy. We had to send our older girls to a store for eggs and bread to eat later after dinner. The pastor's children ate eggs and bread with us, too."

Trying to Farm Again

The Weiss family's sponsor found a place on a nearby farm where they could stay for free, be paid $100 a month to farm, and share the profits from raising pigs and chickens. Immanuel and Johanna jumped at the chance, but, recalled Immanuel, "the pastor or his wife must have called the farmer to say we didn't want it." The same thing happened at a second place and even for an apartment they found in town.

"The pastor was really a good man. He helped our children study for Sunday school, and he was nice in other ways. But they did need us a lot, especially the wife. Three times they called it off. We think it was his wife. We don't want to think evil of them, but it sure wasn't nice for us. We wanted so badly to be on our own, to be independent. You know, maybe God wanted to test us somehow." It seemed the dreams of farming again were impossible.

Finally, Weiss found a farmer who spoke German, so he made the deal in German. Weiss milked twenty cows and took care of 100 beef cattle and sixty pigs. He also did the plowing and much other farm work. A couple of years later, they moved to another farm where Weiss again worked for wages but also shared in the profits. There they had a sixth child. Another year and a half later, they moved to an even better place with running water. "I was no longer just a laborer. I was at least half-independent. Whatever I earned above the rent would be mine." They stayed there six years, and their seventh child was born there.

But Weiss was still not happy. "I wanted to be independent, to have my own farm. To reach that goal quicker, I rented a second farm together with the first. Then in 1959, seven years after coming to America, we found a 250-acre farm we could buy. It cost $40,000, but only $8,000 as down payment." They enlarged that farm and improved the house. Weiss recalled, "This house didn't have running water either, and it was really rundown. But we improved the farm first. A house earns no income!" Clearly, the many trials and sacrifices had only served to energize their intrinsic motivation for independent farming.

How We Did It

"You have to start with old equipment other people don't want. The first year, I bought an old tractor for $70, instead of spending $2,000 for a decent one. I bought an old wagon for $4; I figured it would last me the first year or two. The drag [harrow] cost $5. It was no good, but I figured I'd straighten some old teeth and weld or bolt in some new ones. And so it was usable for a few years. I helped my neighbor plant his corn, and he let me use his planter. For the small grain equipment [drill and combine] and for the corn picker, I worked in exchange for the use of the equipment."

According to Weiss, "The good farmer has to have mental discipline. It takes more than just good land and hard work to make a success of farming. You have to know what's important." He gives the example of this commitment in working with the dairy cows: "Milking has to be done with the right touch, the udder washed with warm water, and the teats slowly prepared with a hand massage. After the milk machine has been on the cow for two or three minutes, you have to check to be sure she isn't milked out. You can ruin a cow by milking too long. She can get an infection easily and even lose her udder. And, it's mostly the best milk cows that need the most attention and careful handling. They're the ones that are the most delicate."

His secrets to success? "First, be healthy. Then you have to keep everybody busy. And we didn't spend money to do certain things, like going on trips or to shows. About the only movies we saw were a few religious ones. We didn't go to dances. After we lost the two farms in Bessarabia and Warthegau, we said that going to a dance means being happy, and we weren't happy yet." In other words, the Weiss family was not going to let extrinsic rewards interfere with the goal of farming. Clearly, they had very strong intrinsic motivation.

As refugees from the madness of Hitler and Stalin, Immanuel and Johanna were happy to have found peace in America. But they never felt completely at ease. They thought many Americans lacked interest in other countries. Immanuel and Johanna eventually visited Communist Romania with a tour group that included some idealistic American college students who had some of their misconceptions about Communism corrected.

Weiss and his family still resented their treatment under Communism and the Americans who had sympathized with the Soviet Union. But most of all they gave thanks to God. "Faith in God and in His guidance made a path for us. Now in old age we can live in peace and freedom in the United States."

8: POSTWAR IMMIGRANTS

After World War II, Americans sent many CARE food packages to their relatives in Swabia. Almost every Swabian family got packages, but not Frau Schäberle. Not that she needed much help, for her family was well-to-do and had preserved its wealth carefully. But she still complained to her maid: "Oh, Anna, if only there had been some good-for-nothings or black sheep in our family, then we could get nice packages from America, too."

Most of the Germans who came to America after World War II were children during the Nazi era. Heinz Biesdorf was a nine-year-old in a well-to-do family when the Nazis came to power in 1933. He disliked his young Nazi peers: "I had to run around with guys who were not my type. I had nothing in common with the masses. I take a shower every day. Why should I hang around with people who don't know soap and water? The Hitler Youth group met only one Sunday a month, but it was too much for me."

Sharing

For the all-day Hitler Youth hikes, Biesdorf's mother would fill his canteen with a quart of tea and lemon. He saved most until lunch. The others started with the same amount, but by lunch they had nothing left. Then they told him he should be "a good comrade" and share his tea. As an individualistic Swabian, Biesdorf was not big on sharing, and he felt the Nazis' attitude about sharing was much like that of the Communists.

In the schools children heard, "Du bist nichts, dein Volk ist alles" (You are nothing, society is everything). The Nazis stressed the *Volk*, the nation or society. Most Hitler Youth activities involved group work, the goal being to develop boys who would one day fight for Germany as a team.[206, 267,p133]

When war later broke out, "unselfish sacrifice" was demanded of Hitler Youth members.[268,p228] Earlier, the Reich Minister of Science, Education and Popular Culture had asserted, "The whole function of education is to create Nazis." These efforts were successful. An American army officer later noted that German boys fought "until they were killed."[160,p119, 206,p18]

According to Biesdorf, the Nazi times were not all bad. While joining Nazi youth groups was still voluntary in 1933, being accepted gave a kid pride. The *Jungvolk* (Young People) for ages ten to thirteen and the *Hitler Jugend* (Hitler Youth) for ages fourteen to eighteen were at first much like the American Cub Scouts and Boy Scouts. The boys hiked, marched, read maps, camped, and learned Morse code. The older boys built gliders and learned to fly them.

However, the introduction to the *Jungvolk* manual stressed duty, and the first section of the book glorified Adolf Hitler's life. Nazi propaganda was also part of the songs and talks in the group. Biesdorf recalled how they heard "the Germans got the short end of the stick in the Versailles Treaty. Nothing else was keeping Germany down."[269]

The Nazi youth groups were not intellectually stimulating for Biesdorf. Hitler asserted, "A man of small intellectual attainment, but physically healthy, is more valuable to the national community than an educated weakling." Similarly, a Hitler Youth leader declared that "we wish to reach the point where the gun rests as securely as the pen in the hand of the boys."[206,pp13,18]

The highest priority in the Hitler Youth was the instilling of physical fitness. Second came character development, especially readiness to sacrifice. Biesdorf heard, "When there is a job you stick to it. It needs to be done, it will be done, and you're going to do it." Only then came scientific education. The humanities were, in Hitler's view, simply aids in cultivating racial and national pride.[268]

Biesdorf's individualism also meant that he was not a good fit with some dissident teenagers who got together outside the Hitler Youth. In Stuttgart and other big cities, middle-class youths gathered in groups to listen to decadent jazz and swing music.

The Nazis hated these youths. Voluntary groups not controlled by the regime were not welcome. Even worse, the tight-laced Nazis hated the "sleazy" teens for their reputed sex outside marriage. The official purpose of sex was to increase the numbers of children who would become Nazis. Swing youths were not actively anti-Nazi, and their numbers were small. Nevertheless, many ended in concentration camps for reeducation and then release to the military.[144,p203]

Social Revolution

Biesdorf liked one thing about the Nazis—they promoted people based on merit. Most Nazi officers were like the elite West Pointers in the United States, people like generals Rommel and Keitel, but some leaders came from the rank-and-file: "Rudel, who destroyed some 500 Russian tanks, was just an ordinary guy. He was drafted, and he started flying the Stuka, and then he knocked out Russian tanks left and right. He got to be colonel."[246,p34]

The Nazis created a social revolution.[270] Hitler's government got people off welfare by paying men to drain swamps and do other hard labor, much like President Roosevelt's Civilian Conservation Corps in the U.S. Between 1933 and 1934, German unemployment dropped from 4.8 million to 2.7 million.[145,p57] According to Biesdorf, "not only did people get money, they were told that they were doing something respectable, not demeaning."

Later, after moving to the United States, Biesdorf felt something similar could help the distressed city of Detroit. He thought he should drive around the city and find unemployed people, ask them how much they earned on welfare, and offer to pay them the same money to paint a house. "Wouldn't that be a good investment? That's how Hitler built the German autobahns."

Germany before the Nazi times had a rigid class system, and the Nazis were successful in giving more opportunities to people in the lower and middle classes. The almost universal institution of the "Heil Hitler" greeting probably also reduced some social differences. When a government employee received ordinary people with that greeting, it could be a sign that the official was the person's comrade and equal.[145] Of course, a few knew the real Hitler greeting was the "forefinger in front of your lips."[271,p22]

At first, Biesdorf recalled, many people supported Hitler because he was a compelling speaker who had a vision of Germany returning to greatness. "Before, Germany was the doormat of every other country. When Germany got that respect, the German people were impressed." When Hitler sent troops into Germany's Rhineland in 1936, demilitarized under the Versailles and Locarno treaties, many Germans felt Hitler was righting a wrong.

Biesdorf stressed that most people did not think about the implications of Hitler's actions: "They just knew they didn't have jobs, and then they had jobs. Everything was much better." When people complained, Biesdorf added, it was usually about local officials or Nazi Party members, not about Hitler. People would say, "Wenn das der Hitler wusste." (If only Hitler knew about it.) Even after the war some Germans didn't blame Hitler for starting the war, only losing it.[145]

Of course, when criticizing officials or Party members, people had to be careful. "People couldn't speak their minds without fearing retribution. They would use what the Swiss called *Der deutsche Blick* [the German glance]. You looked first one way over your shoulder and then the other way."

Biesdorf added, "I'm not trying to excuse Germans. I'm just trying to understand how sixty million people ignored the scope of what Hitler was doing—at least until the war started and German troops started dying. Until then, most people didn't have strong feelings about the regime. But when war came, Hitler's politics were no longer the issue. 'That's us getting killed here.'" Biesdorf likened it to the Russians' ambivalence toward Stalin that later changed to total support, a Holy War, after Germany attacked Russia.

Biesdorf stressed that the Nazis did not have to force Germans to fight to defend themselves. "If you toed the line, there was no problem. Not for the average German."

Under the Nazis, passive resistance was even possible, at least at first. Ten-year-old Biesdorf had enjoyed the *Jungvolk* and playing cops and robbers and other competitive games. But by age thirteen, he started to tune out. He had earlier dropped out of church and gone to museums instead. When his father found out, he reproached him, "You should have told me earlier. I dropped out of church when I was that old, too." So Biesdorf decided he'd also drop out of the *Jungvolk*. Nobody noticed.

When Biesdorf was fourteen, the Hitler Youth was compulsory.[272] He didn't show up for the official transfer ceremony out of the *Jungvolk*. There were no consequences until 1941 when he was seventeen. The war was raging, and there was serious punishment for evading the Hitler Youth, which involved extensive paramilitary training. Fortunately, Biesdorf's dad knew a man whose son was an important leader in the Hitler Youth, which allowed Biesdorf to become legal.

Political Awakening

Biesdorf's political awakening came slowly. His father, an architect, had fought in World War I, and he complained that for Germany's cost of the war every German could have gotten a new house and every Frenchman, too. Then, a high official in the government-run theater, a friend of Biesdorf's father, would tell anti-Nazi jokes and stories when he visited. Later, during the war, he heard his mother saying, "all those countries fighting us now. In World War I, so many were fighting us, too."

In addition, Biesdorf defied the law by listening to the British Broadcasting Corporation despite the penalty being five years in prison, or death. "They never said anything good about the Germans, but when the Germans won a battle, they reported it. That gave them a tremendous credibility." The BBC echoed some of the same dire prospects for Germany as those he heard from his mother.[206]

Then Biesdorf heard a speech by Joseph Goebbels, the Nazi propaganda minister, calling President Roosevelt crazy for saying the U.S. was going to produce 20,000 or 30,000 planes a year. Biesdorf thought, "Why couldn't the Americans do that, if they can produce three million cars?"

After Biesdorf met a medic in the German Air Corps, he decided to attend first aid classes in the Hitler Youth and become a medic. Instead of going on Sunday hikes, he could wait behind to treat the blisters of the others. He also got out of having to go to the boring Wednesday evening meetings. Instead, he attended classes on first aid.

As the war went on, Biesdorf thought, "Those guys [the Germans] are not going to make it." Clearly, by this time he did not see himself as part of the *Volk*, the national community the Nazis emphasized.

Then he heard a friend's mother say, "Wir werden uns zu Todessiegen," [We will keep winning victories till every one of our men is dead]. She added, "We will also lose the war because of the oil which we haven't got."

The Allies had the same idea, launching repeated air attacks on the oil center of Ploesti, Romania, which provided a third of Germany's oil. Success in those attacks eventually destroyed the effectiveness of the German air force and mechanized army, not the nonstop Allied bombing attacks on factories manufacturing the planes and tanks.

Biesdorf decided to avoid the *Arbeitsdienst*, the compulsory work service for eighteen-year-olds that followed the Hitler Youth. It might include digging ditches and other dangerous work on the Russian front. He quit school and apprenticed at tool-and-die work in a factory for six months. Soon the only Germans allowed to work in industry were women, boys of fourteen to sixteen, and men wounded in the war or judged unfit for service.[252]

The draft eventually got Biesdorf even though he weighed only 110 pounds, and he became a medic. If he could get his weight down to 105, Biesdorf would qualify for the lowest military fitness level and a home front assignment. He got down to ninety pounds. He was stationed in Tübingen, a small university town with little industry. It was not likely to be bombed.

When Biesdorf visited Stuttgart, he saw "block after block of skeleton buildings and rubble, and there were people living in those buildings." The Allies dropped 21,016 tons of bombs on Stuttgart. This was about half of the bombs dropped on Berlin, which was many times larger than Stuttgart.[273,p125]

In his avoidance of the army and threats to his life, Biesdorf was only following the examples in the famous Grimm fairy tale of the Seven Swabians. They sought adventures but fled when meeting dangers.

As the war dragged on and Germany suffered more casualties, it became hard for Biesdorf to evade the many extra call-ups, even with his lower weight and his adeptness at exploiting connections.

In February 1943, Goebbels gave his Total War speech saying, "We are all children of our people, forged together." Biesdorf did not agree, especially when Goebbels called on the death penalty for shirkers. Despite the Total War speech, many officials in the Nazi Party evaded the authorized mobilization of everyone. Soldiers newly returning from the Eastern Front would say, "The Russians are conducting Total War, we are fighting an elegant war."[145,p283, 206,p142]

Biesdorf contracted an infectious disease, meningitis. Next, he fractured his skull in an accident that he also considered fortunate, because it saved him from the military for another nine months. It was during this time the so-called *Heldenklaue* (hero-grabber) units checked everyone's papers and sent those without an airtight excuse straight to the front.

As the Allies approached the borders of Germany, Biesdorf's doctor told him, "*Dicke* [fatso], I think we send you home." So on December 19, 1944, Biesdorf was discharged while Hitler was simultaneously ordering Germans of all ages to fight to the death. "This story will not endear me with Germans, whether they were heroes or not. It will definitely not endear me to Americans, because you're supposed to be patriotic and fight for your country. But after the Vietnam War, we're not so sure about that. Before Vietnam, my story used to get me some very dirty looks."

Biesdorf had some positive feelings for the Americans during the war. Nazi propaganda continually harped on how individualistic the capitalistic Americans were. Biesdorf recalled reading a Nazi propaganda brochure, "In God's Own Land," written by someone who had clearly lived in America. "He put everything in there that he thought was very bad, and I thought that's the place I want to be." In America, they didn't care so much about sharing, like in Germany. That social Darwinism appealed to Biesdorf.

Also during the war, he read Henry Ford's *My Life and Work* and was impressed that "workers get no more room at their machine than what they need." Ford described how he planned everything for efficiency. Biesdorf thought that was a good principle.

Biesdorf would later work at encouraging Americans to be more efficient. It was something he had already cultivated while growing up. "My father kept me on very tight reins with spending money, so I played carefully with my Märklin wind-up train set. After eight years I was able to sell it as brand new."

The U. S. was also appealing because Biesdorf hated authority, and he knew he would have a hard time in Germany. Autonomy was important for him. "I was independent from the time I came home from the army." He started his own business in Germany at age twenty-one. He did well, but "I was not geared for Germany any more. I knew the way I wanted to go, and I was stubborn. I wanted to become the most efficient money-making machine, and America was the best place for that." These reasons pushed him to become one of the first German citizens to come to America after the war. His story is continued below, at the end of the chapter.

Schmidt and Working Class Life

During World War II, most younger Swabians had a harder time than Biesdorf, in part because many were committed to fighting as long as they could. In 1944, Hans Schmidt was fourteen and finished with school. Living in a small rural town, he wanted to be a train engineer like his father, who, for health reasons, had been mustered out of the army and was a locomotive driver. Hans began to apprentice at repairing locomotives.

The work took place between bombing raids. They started work at 5 a.m. so they could complete at least three hours before any threat of bombing. If bombers came, they would lie under the railroad cars. They might eat their lunch, "slices of turnip and water." By late afternoon, they were back at work. Children of ten to fourteen didn't have to work more than eight hours and those over fourteen, not more than ten hours a day.[252] Schmidt and his father would get home by 10 p.m. for dinner, either "potatoes and sauerkraut or sauerkraut and potatoes." Even animals had short rations during the war. Boiled potatoes were the family's main food, so their pig had to get by on potato skins, supplemented by husks left over from milling wheat.

Becoming a Patriot

In the *Jungvolk*, Schmidt learned that it was an honor to die defending his country. In the Hitler Youth, he became interested in submarines, wore "a nice blue uniform," and studied the lives of naval heroes. The Marine Hitler Youth had a membership of 62,000 boys, mainly in North Germany, but there were units in Swabia along Lake Constance.[268,p230]

Schmidt's Hitler Youth team competed against other towns in

swimming, track and field, and war games. "They taught us how to compete. It was tough; you had to think for yourself and plan ahead." During the war, there were many cases of boys too young for the Hitler Youth altering their birth certificates so they could join.

Schmidt wanted to help with the war and looked forward to working for *Organisation Todt*, a semi-military construction agency. "At home, we saw only women and old people, while we were young and healthy and could do something."

Schmidt and many of his rural neighbors didn't care about the politics of the Nazis, but they wanted to defend their country once it was at war. "When we were ten or twelve, we saw the [Nazis'] brown uniforms and all that baloney, but when we got to fourteen or so, we only heard about the *Wehrmacht* [armed forces]. After 1943, we heard a lot about the heroic fighting on the Eastern Front, the new lines made to stand up to the Bolsheviks. We had to fight to the last drop of blood. We heard there were secret weapons that would save us."

The October 1944 creation of the *Volkssturm* under command of the SS for men between sixteen and sixty saw even younger boys volunteering.[268]

When Germany was invaded, Schmidt and his comrades figured they'd throw the Allies out. They had been fed hero legends, so they responded to the "*Appell* to the last *Einsatz*" [call to the last battle].[268,p249] They didn't know promises that Germany would push back the invading Allied armies were based on ragtag armies of boys and old men.

A postwar review of internal Nazi reports on German public opinion concluded that "only a small fraction of the teenagers developed a political consciousness: the vast majority blindly obeyed the commands and regulations of the Hitler Youth."[238,p331]

Typical was a very young soldier in the Bavarian countryside in April 1945, weeks before Germany's defeat, who asserted Hitler still promised victory, "and he has never lied before. In Hitler I believe. Why, God will not leave him in the lurch! In Hitler I believe."[239,pp270-71]

Schmidt recalled, "We thought we could assemble in the mountains in Austria. We figured the generals were going to take a new approach. We knew the Gestapo was bad, but we figured that maybe we fight in a different way. Maybe we team up with the Americans against the Russians, not Nazi style, not like the Gestapo, but fighting honestly, if you can say that. But nothing happened."

The Bombing

Most German civilians were also stubborn in putting up with the terrible bombing of their cities.[274] The Allies often used carpet bombing, dropping thousands of bombs continuously across cities. The densely-populated city centers were often easy "aiming points" for raids.[275,p265]

Ironically, surveys after the war showed that German industry, the declared target of the bombing, continued through most of the war relatively unaffected. The eventual shift to strategic bombing focusing on particular industrial sites had little more effect. In Germany as a whole only 15 to 20 percent of industry was wiped out in the air raids.[257,p125] Personnel and equipment were usually shifted either temporarily or permanently to new locations. The destruction of non-industrial sites also ironically freed up many former non-essential workers, such as waiters and shop clerks, to be absorbed into the armaments industry.

An added goal of the Allied bombing was to damage the morale of the civilian population, although American survey research in Germany after the war showed that civilian spirits did not drop much after cities had received only a moderate tonnage of bombs.[276,p23]

It is hard to defend killing civilians to decrease morale. Bombing, according to American scientist Freeman Dyson, "made evil anonymous. Through science and technology, evil is organized bureaucratically so that no individual below the very top is responsible for what happens."[277,p24]

Study showed that it took two and a half tons of bombs to kill a single German. Was the bombing campaign worth the 79,265 American and 79,281 British aviators killed?[206,p191, 278,pp331-32]

The bombing would eventually persuade many Germans that they were the victims of the war. By the last two years of the war, when there was intensive bombing and much resulting hardship, the Jews were only a distant memory for most civilians. If Germans did know about Babi Yar or other murders in the East, and some did, the bombing perhaps served to dissolve their guilt since they could now see themselves as victims.[145]

Eva Bromke lived under the threat of bombing as a child. Augsburg in Bavarian Swabia was an important industrial center, but there were few attacks until late in the war. Allied fighter planes at first did not have enough range to escort bombers that deep into Germany. But while growing up, Bromke saw refugees who had fled bombing raids on the big cities further north.

For two or three years, she and her mother heard air raid sirens, saw searchlights sweep the sky, and sometimes heard planes in the distance, but nothing more. Then one night she was frightened by sudden booming noises that shook her building. These were nearby flak or anti-aircraft guns. There were 20,693 spaced throughout the western part of Germany. Some could hurl a 57-pound high-explosive shell to an altitude of 35,000 feet. These fearsome guns had shot down 8,706 planes by December 1942, but Allied fliers carried on.[206,p90]

Eva and her mother lived on the second floor of their apartment building. They would always have at the ready suitcases packed with essentials—clothes, jewelry, important papers, and some food like crackers and salami—to take to the basement when bombing threatened.

Radios were turned on non-stop, and at night people spelled one another listening. There was normally a monotonous *tock-tock* sound from the radio, but when it changed to an insistent *ping-ping*, it meant enemy planes had entered Germany. An announcer would give the position, heading, number, and kind of aircraft.[206]

When the threat of bombing was broadcast on the radio, all six of the families in Bromke's building would take their packed supplies and valuables to the basement and wait on cots, sofas, or bunk beds until the all clear. They would be able to stay there for a whole day or even evacuate to the suburbs directly from the basement.

If air raid alarms lasted past midnight, Eva's school would start later, at 10 a.m. "If it was a very long night of alarms, there was no school the next day." During daytime alarms, she and other children sheltered in the school's basement.

After a while, some Augsburg residents did not wait for the air raid alarms. If they had no basement shelter, at dark they would line up at public shelters to be sure they got a place.

Augsburg's luck finally ran out in February 1944. Several times, bombers came in during daylight, flying low and evading radar to target the Messerschmitt airplane factory. The fourth raid used carpet bombing, since strategic bombing wasn't yet accurate given the strength of flak and fighter defenses. That raid by 594 British bombers was unforgettable for eleven-year-old Eva: two-and-a-half hours of continuous carpet bombing. After an hour's pause, two more hours of bombing. "They were such big bombs that we couldn't catch our breath. They sucked the air from our lungs."

One account describes a group of Hitler Youth officials and their girlfriends in a public shelter: "At first they were noisy and insolent, then, as things got worse, they knelt in a circle on the floor, clutched at one another and ducked every time there was a heavy explosion. In the end they were praying."[206,p144]

After the all clear, Eva and her mother climbed up through the rubble on the staircase to their second-floor apartment. "No more windows, curtains, lamps. The bigger things like beds were still there but covered with dirt and debris." The explosions had moved the six-story building off its foundation by a foot and a half. One wall was completely gone. That same night, her grandmother's house was totally destroyed, but she survived in the cellar with one suitcase.

Fleeing Augsburg

Despite the bitterly cold February night, the air was heated by an inferno. "Augsburg was burning and seemed practically leveled to the ground." This resulted from a change in Allied tactics. A 1942 British study showed that one ton of high explosives destroyed 2,000 square kilometers of cities while one ton of incendiary bombs destroyed 13,000 square kilometers. An American study similarly showed that attacks with explosive bombs plus incendiary bombs were preferable, five or more times better than explosive bombs alone. The earlier German attacks on England had not been so potent since the Germans hadn't yet become aware of the advantages of bombing to create fires.[273,p99]

Part of a great exodus, Eva and her mother walked for several hours through the rubble to a small railroad station at Gersthofen, where the trains still ran. Along the roads, farmers brought tea from their homes to the people fleeing the city. Eva and her mother were part of the fifteen million Germans who eventually fled the cities to escape the bombings.[259,p51]

Many did remain in the cities to continue working at their jobs. The surviving Germans were astoundingly resilient in clearing away the rubble, restoring utilities and transport, and continuing their daily lives. Governmental bureaucratization and the preference by the people for continuing their routines, as well as accepting authority and duty, fostered this resilience.[278]

Eva and her mother had no reason to stay in Augsburg, so they moved in with relatives in the country. Over several years, Eva's mother had methodically stored beds and bedding, kitchenware, and her more valuable porcelain there.

Eva was so terrorized by the bombing that she vowed never again to go into Augsburg. Yet she did, three times, including once for a burial, and there was bombing each time.

There were 2,850 tons of bombs dropped on Augsburg between February and July of 1944. Allied analysis showed that 29 percent of housing was destroyed in those few months. The ostensible aim of the bombing was to reduce productivity through absenteeism and damage to industry. However, five months after this series of raids, 80 percent of the former productivity in Augsburg had been regained.[274,p73, 275,p266]

Some of the continued productivity was due to the use of forced foreign workers, including prisoners of war. By the spring of 1943, there were twelve million in Germany. They comprised 40 percent of the nation's workforce, and in some armament factories 90 percent of the workers were non-German. They couldn't flee the bombing like Eva and her mother.[206,p142]

Rural life was not idyllic. The farmers left in the dark of early morning to start cutting grass for the cows. Later in the day, Eva would take the horse and wagon to haul the grass back to the barn, only to have the low-flying British fighter planes shoot at her and other children. The planes were so low she could see the pilots shooting at them. Such strafing was widespread. According to a 1944 German secret-service report, female farm workers were refusing to go into the fields.

Another time when Eva was on a train, everyone was ordered out. They lay in the grass as far from the train as possible.[206]

Most Germans understood that their country was beaten, and they wanted an end to the war, but others couldn't see themselves living in defeat and occupation. A joke circulated in late 1944 and 1945: "Kids, enjoy the war! Who knows what peace will look like!"[145,p290]

This mentality to continue resisting is suggested by the reaction to the failed assassination attempt on Hitler on July 20, 1944. Many saw it as another "stab in the back." There were massive demonstrations in support of Hitler, with 350,000 people at an outdoor rally in Vienna. Most people could not imagine a future without Hitler. However, the last ninety-eight days of the war in 1945 cost the lives of 1.4 million soldiers.[145,p291]

For a whole week, Eva and her mother saw wounded and starving German troops limping through their village, with horses pulling the remains of their equipment. SS men then came in a car with leaflets telling people to fight to the death because the American conquerors were raping women and children and killing every man left.

Eva thought she had only a few more days to live and stayed close to her mother. "We looked at each other, and we were very quiet. We could hardly talk anymore, just waiting for what was to come."

Two days later, they saw "two fanatic village boys. They were eighteen years old, in SS uniforms, and they carried a *Panzerfaust* [anti-tank rocket launcher]. The boys lay in a ditch at the edge of town." They probably didn't think of surrender because they had been told so often about the odious "stab in the back" at the end of World War I. A few old men and some women took pitchforks and hammers and told them the village would not fight. "We will hang the white flag out. Then we will see. Whatever will be, will be."

Eva and her mother saw the American tanks approaching on the road, with infantry to their left and right. "There we stood, behind a window, shaking like leaves, and quiet like in church."

Enemy Soldiers

Starting at the edge of the village, a team of four soldiers stopped to check inside each house. After the first three houses, they came to Eva's house, but with ten men. They took their time and searched everything, mattresses off beds, all cupboard doors opened. It was a thorough search. To her surprise, one of the soldiers gave her a piece of chewing gum. "Ugh," she thought. "What awful candy."

The soldiers still would not leave. Eva could not understand what they were saying. She asked again and again what they wanted. Finally, she understood what was wrong. They asked Eva, who knew a little English, why they did not have a white sign of surrender like other houses in the village. She blurted out that they were not hiding anything. They just had not thought of a white flag. The soldiers politely left.

Like Bromke's small village, Augsburg also surrendered to American troops without a fight. The city was almost half-destroyed, but Berlin had ordered they fight to the last man. However, the mayor and the resistance took prisoner the commanding general who still believed that orders were orders. On April 27, the city had gotten a message from the U.S. Third Division threatening an attack of 20,000 bombers in a half hour unless there was a surrender. The city was taken with little further damage and little loss of life. However, those who arranged the surrender were arrested.[279,p272]

After Bromke's scary experience in their small village, she recalled that "the American occupation soldiers showed no resentment, no hostility towards the German people. How disciplined these troops were. This was said all over Germany. The first occupation troops were so very different from the ones that followed later."[280]

Bromke recalled, "People were glad the war was over. Not everybody was a Nazi. People used to say the war was a lost cause. They were hoping something would happen so there would be a surrender, but it never came because of the SS and because people were convinced there was no way to protect their families."

There was fear: "We knew there was a concentration camp. People would tell others to shut their mouths and not say anything about the system because you'll be taken to Dachau. There was a man in our neighborhood, a Social Democrat [socialist], and he openly opposed Hitler in talking to customers at his bakery. He was warned, but they eventually took him away to Dachau. Then his wife got a letter that he died of a heart attack."

After 1936, anyone released from a concentration camp had been warned, under penalty of death, not to say anything about what they had suffered there. This silence added to the fears among those who were unhappy under the Nazis.[206]

"I also saw concentration camp inmates with stupid suits on who were repairing the trolley car tracks. We didn't know if they were Jews or what they were. The newspapers would say they are criminals or enemies to our system." According to Bromke, "I knew as a child that there were people that were pro-Hitler, but there were also people who were anti-Hitler." The latter was more common in

Catholic areas like Augsburg than in Protestant areas.

The Military Occupation

Eva returned to live in Augsburg with her divorced father in 1945, and she then moved to live with her mother after their apartment building was rebuilt three years later. However, the occupation and military government made life difficult for many in Augsburg.[279]

Members of the military government had been trained to get communications and transport going again. There had been no preparation for military governance of an occupation. The troops were soon young replacements fresh from the States, with just a few career professional officers. They were asked to accomplish tasks they apparently did not care much about. When the Army sent home the wartime troop strength in 1945 and 1946, it removed most of a professional force.

There was plundering in Augsburg by American soldiers, and when the German police tried to stop them, several police were killed. Even high officials who had not been Nazi Party members were arrested, while Nazis were allowed to take some positions in the government. Forty three of the forty eight doctors at hospitals were fired, as well as fifty eight of the sixty firemen and nearly all the policemen. The denazification program dragged on slowly. Finally in 1947, instead of former Nazis being put on trial, they were just given a fine. A thorough U.S. review suggested that "the denazification program for one was ill-conceived and literally impossible of execution."[279,p276, 281,p357]

People in Augsburg suffered from the occupation in various ways. The thousands of former slave laborers, freed by the Army, looted homes in an agony of reprisal. Women were being molested by soldiers so that those who were employed by the occupation started wearing an America flag for protection. German women caught on the streets could be unceremoniously hauled in for the Army's venereal disease examination.

Homes and businesses were seized for officers' housing with only hours warning. When American families arrived in 1946, still more houses were confiscated. Foreign workers from Lithuania, Estonia, and Poland were given housing, but many had been Nazis and even SS. The food crisis reached its peak in 1948. There was continual waste and even burning of food by the Army in front of starving Germans. The Army did not return the last houses to former German owners until 1957.

Eva recalled that as the occupation continued there were "definitely incidents." Americans were loud and impolite. They did not behave with the decorum that Germans expected. They played baseball in the streets. Soldiers did not interact with the Germans unless it was for sex or for black marketeering.

While some soldiers were giving children and women food, other soldiers were robbing adults. Members of the resistance, the first to see the Americans, were the first of many to lose cars, watches, and wedding rings. Robberies by soldiers became more serious after the currency reform of 1948 because soldiers could no longer engage in black marketeering. Before, soldiers could go to the post exchange and buy a carton of cigarettes for a dollar and trade it for $200 of goods from Germans on the black market.[280,p155]

Eva had one frightening experience. One day in 1946, she crossed the bridge over the River Lech to Augsburg's eastern suburbs where her grandmother lived. American sentries were at both ends of the bridge to check Germans for their *Kennkarte*, the identity card that everyone had to carry. She was only fourteen, still a girl in German eyes, but "already such that a GI would not let me go." He told her to wait. She was horrified. She knew that they sometimes took young women away in their Military Police trucks. She stood there for an hour, getting more and more nervous.

Her father worked for a large grocery chain and used a car with "Military Government" on the door. He lived across the river in the Augsburg suburbs, and he happened to drive by. He asked Eva, "What is wrong here. What is the matter?" She was close to tears as she explained, "I'm just standing here. I don't know why." Her father simply told the GI that he wanted to speak to his commanding officer, and that solved the problem.

Most younger Germans, Eva said, were friendly towards the American soldiers. "But there was the older generation, the diehards and never-give-ups." Some of them, she said, had been "totally indoctrinated by Hitler, so brainwashed, that they were beyond reach. They would complain how all the wild things came from America, all the crazy music, all the bad manners, so many uncultured things." She knew the feelings, for she used to have some of that same arrogance. An example of the culture conflict involved the only big auditorium in the city. At least as late as 1949, the Americans used it for boxing matches instead of theater and operas as the mayor wished.[279]

It is said there was no lasting harm in the collision between the two cultures because of two factors. First, there was a continuing effort to keep a soldier within an American environment. His duty, recreation, radio listening, newspaper, and even physical surroundings were American. If he ventured out into Germany, it was rarely, and only as a tourist.

Second was the influence of the German women. They were more accessible than in America. It was easy to pick up German women. They gathered around most garrisons. Some were out-and-out prostitutes, while others sought a more lasting relationship. Soldiers learned rapidly that "German women were attractive, buxom and willing—after all, Hitler had preached a certain sexual license." These contacts had cultural and psychological impacts on the occupation soldier, with some resulting in marriage. During 1945-1949 in the American occupation zone an estimated 10,000 German women became "war brides."[280,pp 145,117]

A War Bride

Eva did not come to America out of economic necessity like so many other Swabians. She fell in love with an American soldier named Chester Bromke.

Many war brides were working-class women or refugees, or they had family connections in America.[281] Eva was middle class and had no relatives in America. She had been unlucky in love, losing one boyfriend in an air crash, another to his former love, and a third because his Italian mother disliked Germans.

A girlfriend invited Eva to a party at the American officer's club in Augsburg. "I'll go once," she thought. She wanted to speak English and find out what the Americans were like. There she met Chester.

Germans looked down on women who dated American GIs, so Eva tried not to be seen with her beau during their courtship. Their dates were usually to movies on the post and to the officer's club. They knew each other for three years before they married.

Impressions of America

When Chester was reassigned stateside to Fort Campbell, Kentucky, Eva's German friends warned that she would be unhappy. She had lived a cultured life in Augsburg. Friends told her to stay in Philadelphia, where her husband's mother lived. But when Eva arrived in Philadelphia in 1961, she was struck by how ugly it was. "Downtown Philadelphia later became really elegant and beautiful, much more to my liking, but when I came in 1961, I was disappointed."

When Eva and Chester walked around downtown Philadelphia the first day, she found herself wanting desperately to sit down at a cafe. This was part of the cultured life she had experienced in Augsburg. "We would take our coffee breaks from work, sit for half an hour, smoke a cigarette, and meet some friends, or just sit and watch dressed-up people walking by. Almost everyone with a business on our street would sit there."

Down at the docks where they picked up their car, she saw "so many *clochards* [bums] like I never saw before, not even in Paris along the Seine. I was afraid in that town, and I wanted Chester to take me for a coffee NOW, to sit down at a sidewalk cafe. But there was no such thing."

When they drove to Kentucky, she thought: "What a beautiful country, not overcrowded like in Germany. There our resort lakes were always crowded with people."

Nearing ten p.m., after driving all day, they stopped at a little restaurant. To her dismay, it did not even have tablecloths. She ordered a beer, but then the waitress almost immediately took it away. It was illegal to drink after ten p.m. She thought, "What a stupid attitude."

Young Germans like Eva used to hear so much about the opportunities and the tolerance in America. "But then you are disappointed, like the rules for alcohol."

But she did miss Germany's class system where people could be comfortable socializing with their own kind. She also missed German socialized health care. She had to be more careful to save money for emergencies. In America's free-enterprise system, she had to watch out financially so much more. You can "better yourself a lot in America, but when you fail, you are also on your own." As a result, she felt "less protected." She became cautious, thinking carefully about what things would cost.

Bromke did enjoy the greater personal freedom in America. She found there were fewer expectations and requirements for social behavior in America. She was learning that American culture is loose, not tight.[2] "When I visit Germany, I have to greet all my relatives. If I didn't come and drink coffee with every darn aunt, I would commit a major offence. I would be regarded as an unmannered person." In America, she felt free to say to her mother-in-law, "This week we are busy, we're coming to see you next month." It would not be taken as an insult. "People are freer here and more understanding. The old traditions in Germany can become a burden."

Bromke also came to appreciate Americans' tolerance of foreigners. She eventually moved to New Jersey, where "there is no such thing as a foreigner or a stranger. I don't think there is another place in the world where people are really so integrated. America is far advanced in that respect." Bromke felt immigrants should not judge America until they've experienced the country for five years.

Bromke Speaks Out

Already as a preteen during the war, Eva was warned to avoid asking political questions. But she later became outspoken as an adult. For a while, her husband was stationed in France with NATO, and there she defended Germans against French friends. When visiting Germany, she defended her French friends. She later did the same with her American friends.

In America, Bromke was also able to question authority in a way that would seem foolhardy to many Germans. She challenged her child's schoolteacher, which is rarely done in Germany. She accused a Jewish schoolteacher of prejudice in a New Jersey school system and community where Jews were in the majority.

The problem developed in the early 1970s when her husband, in full military uniform, conferred with their son's schoolteacher before going off to Vietnam. After his visit, their son's grades suddenly dropped, and the child reported the teacher was hard on him. The teacher insisted the child was "not attentive" and had "a real stubborn head, a typical German stubbornness." She blamed the student's bad behavior on his father going to Vietnam, a war the teacher opposed. Bromke retorted, "What would you say if I tell you that you have a typical bad Jewish attitude," and she threatened to tell the principal about the teacher's prejudice. According to Bromke, "I do not want my child having to pay for the Holocaust."

Stubborn Questioning of Religion

Eva Bromke displayed an unusual stubbornness from childhood well into adulthood as she tried to sort out her religious and spiritual heritage. She considered herself Catholic, even though she didn't attend church every Sunday like her husband. However, even as a child she had wondered about Catholic teachings. How could all the different races come from Adam and Eve? Was the Immaculate

Conception possible? How can an innocent six-year-old have Original Sin?

In her school in Germany, the priest had taught a couple of hours of religion every week, as is true elsewhere in Germany. He gave grades in it, just like any other class. Such extrinsic rewards could have been a disaster for her spirituality, but Bromke recalled the priest as intelligent and good. He showed tolerance and "lived his religion." He was open to discussion. As Bromke saw it, "It is two different things—the dogma of Catholicism and to grow up with good people who happen to be Catholic." This was the relatedness that is so critical in fostering intrinsic motivation.

She read widely, "turning many pages," as she stubbornly tried to come to terms with the beliefs of the Catholic Church. She eventually agreed with writer Hermann Hesse and his view that people who cannot think for themselves need the crutches of a dogma. She saw religions as "political parties. They govern people. They care for them. They have to receive people who are unhappy, who cannot by themselves accept life as it is. There are people who always think 'why me, why must I suffer?' They have to believe in something. They need a figure and even a holiness created for them."

She considered: "Maybe this is oversimplified. I don't want to be intolerant." She paraphrased Solzhenitsyn, "Who is to tell right or wrong? The real, holy conscience is the only definition."[282] To that, she added a warning from Hesse that humans should not make too much of their importance. "God is concerned no more with humanity than with earthworms. We falsely believe we have an enormous power to control our destiny, and we get fanatic about this way or that way to God." She insisted, like Hesse, that people of all religions should get together, and not be so fanatic about their particular God.

She eventually concluded that religion was a spiritual and ethical matter for the individual to decide. "Once religion becomes organized and concerns itself with accumulating wealth and power, it is hypocritical when it speaks of love."

Bromke's reading of Hesse made her give up her stubborn attempt to understand religion. He made her finally understand it, especially his book about Siddhartha. Teachers in India told Siddhartha to give up all pleasures in life, but he finally realized that meaning comes from the process of life, in living.

"I had an insatiable curiosity, but now I am contented in that respect. I experienced life and realized there is no one on earth that knows the unknown, knows the ultimate answers." She gave up her puzzling over the religion of her childhood. She came to terms with it, but on her terms, not those of Catholicism or of any other organized religion. She came to see there was a place for her commitment to the Catholic religion of her upbringing. Her stubborn search finally allowed both contentment and the preservation of her religious feelings and her spirituality.

Welfare from Hitler

Swabians often combine their love of work with their love of independence. They become entrepreneurs. This was the route Rudy Geiger took to become a millionaire.

He was born in 1929 into a poor family in a Black Forest village near Rottenburg. They had only six-acres of land and two cows. Geiger's father worked at odd jobs, in the village woodlot or doing haying for bigger farmers. His mother did the farm work: milking, feeding the pig and chickens, growing potatoes, vegetables, and wheat, and gathering hay. She even did some of the plowing.

Geiger remembered how his family got welfare "thanks to Hitler." In winter, they would go to the village office to collect coupons for coal and clothing so they could survive. "Winter Help" was a nationwide program to stress how the Nazis were creating a community of all the people. The Nazis asserted that "a people helps itself."[145,p53] However, villagers looked down on people like Geiger's family that got welfare.

Geiger and his family were lucky since they were not considered candidates for the 1934 "Law for the Prevention of Genetically

Damaged Offspring." Candidates for sterilization came from people in asylums and hospitals and on welfare rolls. Later there was also a euthanasia program for mentally ill and handicapped people, but the outcry by the Catholic and Protestant churches halted it.[283]

Geiger recalled, "I had no life." He went to school, and after school he did chores. "If not for my family, I worked for someone else." He lamented never learning to swim like the other kids. He never had the free time. After elementary school, he apprenticed as a carpenter.

Geiger's Turning Point

One Sunday afternoon in the village, he and his father heard music as they passed the local inn. They thought they would go in to get a beer. But when his father looked in his pockets, he couldn't find even ten pfennig (two and a half cents).

Geiger vowed then and there that he'd eventually make so much money that he could buy anything he wanted. He has told this story to perhaps a thousand people, because it was a key event in his life. He stressed, "You have to have willpower inside. If you want to be successful, you can't relax. That's why today, there's nothing in the world that I can't buy."

Geiger was lucky to make it through World War II intact. Like other fifteen-year-old boys, he was a member of the Hitler Youth, eventually becoming the leader of a small group. He believed in doing his duty and fighting for his country, and at war's end he was guarding a bridge over the Rhine between France and Germany. His unit surrendered, so he was only beaten by the French and imprisoned.

Geiger and his wife arrived in America in 1957 with $1,000 in savings. They had given up good jobs in Germany. His first work in America was as a sweeper in a clothing factory, but he had other part-time jobs. He would clean a church before Sunday services. There he would sometimes pick up pennies from the floor. At a parochial school he did carpentry and plastering in the evenings. He worked at least ten or twelve hours every day, including weekends and holidays. Most of the money earned was saved. "A salami used to cost fifty cents, and from that salami, we ate all week."

As part of his job at the clothing factory, Geiger unloaded trucks while the driver would go inside and have coffee. One day, a driver noticed his positive work attitude and told him he had a friend who had capital for a business but needed a working partner. So Geiger began installing aluminum siding on houses while the partner mostly worked on sales, since he was an American and spoke English well.

At first, Geiger's partner figured out the expenses, including the materials he bought, and the two would split the remaining money. "I wouldn't really know what the costs were. My partner probably made more money than me." Geiger got more involved in buying the materials, and he also brought in other workers and paid them himself.

An Entrepreneur

After two years, Geiger decided to go into business for himself. He met the owner of a dairy business that was expanding from a couple of small retail outlets into a chain of convenience stores. The owner was of German-Russian descent, and they hit it off. Eventually, Geiger had fourteen men working for him siding grocery stores and houses. It was not always easy. "I don't know how many times I taught the business to Germans, and then they left. Only the dumb ones stay with you."

Geiger felt that because he learned his trade in Germany, he was more conscientious and did better work than Americans. "The American-born, when they are done for the day siding a house or building, they will leave things lying around. I had my men quit at ten minutes after four, sweep the driveway, rake the grass, and, in general, clean up the job. They did that every day until the job was finished." He also had high standards on trimming and fitting the siding. Once, he had the men tear down a wall after he found it not done right. He lost money on that job, but he gained a reputation for good work.

Geiger was also frugal in using materials. "I knew exactly where a leftover piece could go, like between two windows, next to a chimney, or on a gable. There was no waste. Other builders didn't care. There was one builder, after he cleaned up his jobs, he left two-by-four and two-by-eight lumber and big pieces of plywood. I'd pick them up, take the nails out, and use them. That was new wood!"

Geiger worked weekends and some holidays. If he went to a party on Sunday, he was home by 6 p.m. and in bed by 8 or 9 p.m. "Even if I didn't sleep, I rested." He'd often lie awake until 2 or 3 a.m. "My mind kept working. I'd be thinking about the next day. I wanted to know exactly what each person had to do so there was no wasted time."

Geiger tried to slow down and reduce his work hours. No more working until nine or ten at night during the week. He even gave one of the men some of his contacts and equipment to work on his own. "But the man didn't want the responsibility. He was a worker, not a businessman."

Geiger's workaholism finally had some negative effects. Geiger came down with a variety of ailments, but the doctor could find nothing wrong. He kidded Geiger that maybe he belonged in the state mental hospital. Even though Geiger had no real financial or business worries, he often could not fall asleep at night, and he would break out in cold sweats.

Geiger eventually dealt with some of his stubborn work ethic. While he at first saw work as a way to get more and more money and avoid the poverty of his childhood, he came to stress the intrinsic satisfaction from jobs well done. He would do jobs for less money than he could have gotten. He was happy doing a good job and then simply charging a fair price. The extrinsic reward of money couldn't overwhelm the intrinsic satisfaction of quality workmanship. Of course, by that time Geiger was a multimillionaire, so money didn't mean so much anyway.

Making Americans Efficient

After Heinz Biesdorf came to the America of his longing, he started working at a Heinz factory in Pittsburgh on a can machine. He added a night job at a hospital laboratory, all the while going to college, first at Duquesne then Pitt. He eventually became a professor of consumer economics at Cornell University.

As a believer in efficiency, Biesdorf devised a "Super Shopper" program for consumers. It was based on his personal practices and those of his Swabian wife, Ellen. Biesdorf defined efficiency as getting more out of the same money. He also taught the notion of opportunity cost. "Whatever you do, you'll miss out on something else. If you spend extra money for something, for example something not on sale, you can't use that money for going to Paris."

Biesdorf focused on grocery shopping. People do it every week, so there are many chances for learning, for getting feedback and correcting errors. Biesdorf felt the techniques for efficiency at grocery shopping could be applied to other areas of consumption, like buying clothing, buying and operating an auto, even a house. Biesdorf believed that consumers could easily save a tenth of the two million dollars they would spend in a lifetime, or enough to buy a house, if they shopped correctly.

The Super Shopper Plan

Biesdorf's Super Shopper program was simple: First, buy at the right price. Get to know when items are on sale. Then buy in quantity, rather than just enough to last until the next shopping trip. When he'd lecture, he would tell people to buy extra green bananas if they were on sale. "You're not going to die before they get ripe."

He would then give another example with peanut butter. One might normally buy a jar of peanut butter once a month for $2.50. When it is on special, it's only $1.50. That is when the Super Shopper should buy five or ten jars. Then, every so often when a jar of peanut butter was empty, there was another jar in the Super-Shopper's home pantry at 40 percent lower than the usual price.[284]

The Biesdorf family also saved some money in the summer by planting a small vegetable garden in the landscaping around their urban home. But they saved even more money all year by Ellen Biesdorf's cooking. She made many things from scratch, despite working fulltime as a nurse. The Biesdorfs did not buy prepared and processed foods, like cake mixes, frozen dinners, and the like. These did take time, but "you don't calculate how much you save per hour. They aren't work. Do you put a price on time spent fishing, bowling, watching television, or anything else you enjoy?"

The Biesdorfs also used leftovers. Swabians would never throw out edible food. This can be a major source of savings. A University of Arizona project calculated that 15 percent of the food entering Tucson homes was wasted. An extra 10 percent of food went down garbage disposals. Another study came up with a similar estimate of 26 percent waste.[285]

Biesdorf also applied his principle of "never buy at the regular price" to shopping for clothing and other items. He knew when certain items went on sale, for example linens in January. They would stock up then. Men's white dress shirts will always be in style, and Biesdorf went through a lot in a year, so if he saw them at a closeout sale, he would buy plenty.

Biesdorf admitted the Super Shopper plan required time. But half an hour a week could pay off more than what a person earned in the same time on the job. "Besides," added Biesdorf, "the government doesn't tax the money you save."

What may be harder to quantify is the psychic energy needed for shopping. One cannot be an impulsive buyer. One has to resist temptations. But for the Biesdorfs, this was all a matter of habit. It might be hard at first and there would be mistakes, but after a while it would be an automatic commitment. "My wife and I have certain things that we just don't do. We're at the mall frequently, but if we're thirsty, we won't buy a drink at a restaurant, because in five minutes we can be home. It's like honesty in children. Instead of learning don't steal this and don't steal that, children learn not to steal in general. The children never even think about stealing. So it is with us. We don't even think about a restaurant."

In Biesdorf's view, most Americans do not plan enough. They do not think seriously about the future. They should be ready for opportunities. That is how the Biesdorfs bought their house. They had thought about their needs, studied the market, and lined up financing. When something became available at the right price, they snapped it up, in one day.

The Biesdorfs looked for efficiency in other parts of their lives, too. Biesdorf thought about underproductive time and tried to make it more productive. "I might feel real tired at the end of the workday, but I can't go to sleep then, because I'd wake up at 3 a.m. So I use the time for shopping. It has to be done anyway. Someone else might watch television, but I will go shopping or I'll spend time in my garden. Either way, I'll get something out of it. The other guy gets nothing."

Biesdorf stressed that once you have money saved up, you are living on a different level. "You should never pay someone 18 percent on your credit card to buy something that wears out. By buying on credit, you wind up paying double the cost of the item. The same goes for buying a car on credit. When you have savings, you can get a newer car that needs fewer repairs. Plus, you have the money to keep it well-maintained and prevent repair problems. Having savings and not having savings is like the difference between a guy who has two legs and a guy who has only one."

On a Mission

Biesdorf had a mission. Like other people in this book, he was dedicated to a principle. While he believed in efficiency for himself, he wanted to help other people become efficient—as many people as possible. He contrasted his work to a missionary in Africa who might spend a couple of months converting just two heathens. If Biesdorf presented to a group of even fifty people, he would think, "There are twenty million people in the state." That is why he focused his extension work on the mass media. He thought, "If I help a million people save five cents each, that's $50,000, much better than if I help twenty people save a dollar each."[286]

The Super Shopper program eventually became so popular that Biesdorf appeared on many television shows and was featured in *Money Magazine*, *Family Circle*, and *Parade Magazine*. After he did a program on WOR, the largest radio station in New York City, he was asked to return for twenty seven more appearances. He sent his Super Shopper program materials to sixty newspapers in the state and allowed them to use excerpts. He also sold several thousand presentation packages for educators, so millions were eventually reached with Biesdorf's ideas of living efficiently. This was amazing considering that Americans, in the 1980s and 1990s, were more well-off than they had ever been before.

The efficient lifestyle of the Biesdorf family paid off so handsomely that they eventually earned more money from their savings and investments than they did from their two jobs. They finally realized that they would never spend down their savings. They had to try harder to spend more money.

However, their intrinsic motivation for efficiency persisted. "For example," Biesdorf explained, "In twenty years I've eaten out for lunch maybe three times. I'd rather take a sandwich to work."

9: CONCLUSIONS

Development of Intrinsic Commitments

We have seen three important components involved in developing intrinsic motivation and the resulting single-minded commitments: (1) autonomy or freely chosen action, (2) a feeling of competence or efficacy from the behaviors, and (3) a feeling of relatedness with the other people involved.

In addition, if one finds that freely chosen actions produce negative results and accepts them, the commitment can become even stronger. The uncertainties, alternatives done without, and sacrifices accepted are all relevant. Integrating or bonding these negatives with the positives, an emotional working through of conflicting feelings, may yield a new and stronger synthesis. Committed people actively build and maintain positive illusions and ignore negative information to preserve their commitment.[30]

Conrad Weiser (Chapter 1) freely accepted a missionary's proposition that he live among the Mohawks. Such freedom of choice is essential in fostering internally rewarding feelings. No doubt, he had some feelings of competence as he mastered the Mohawk language.

The early development of Weiser's commitment to friendship between the Indians and colonists must have started with the positive reception, the relatedness, from the Mohawk chief and his family. Perhaps that reception was seen as positive because Weiser was escaping from a harsh stepmother as well as a father that later accounts describe as a difficult man. Weiser probably also compared his friendly treatment among the Indians with the bad treatment he suffered at the hands of the English at Livingston Manor and later when the English came to Weisersdorf to take away the lands granted by the Indians. Research also suggests that even the simple repeated contacts with the Indians helped foster some feelings of attraction.[287]

Weiser admitted that he had to hide from drunken Indians. He apparently rationalized and accepted this downside. He later remarked that he never saw any violence among the Indians apart from that fueled by alcohol. He may have recognized how the Indians were superior to the colonists in this respect. If his negative experience with drunken Indians had occurred when he first arrived, he might have left. But once he started to have some good feelings for the Mohawks, the bonding of negative and positive experiences together promoted the intrinsic motivation and commitment that lasted his entire life.

The commune members in Chapter 2 already had a strong religious commitment while in Swabia. The Separatists underwent many negative experiences there that must have strengthened their commitment. The Lutheran authorities placed many restrictions on the assembly of the Separatists. Some of them were imprisoned. No doubt, the difficulties in escaping from Swabia and traveling to America further strengthened their intrinsic motivation to await the Second Coming predicted by Father Rapp. As we pointed out earlier, the serious sacrifices at Harmony commune fostered such strong convictions that Harmony lasted 101 years—longer than any of the thirty other communes that existed during the nineteenth-century.[37]

Zoar commune was not as strict as Harmony. Sexual relations were eventually allowed and so were entrepreneurial activities and outside contacts. The intrinsic commitment to the commune had to compete with extrinsic rewards. The commune's eventual demise occurred when an outsider introduced further extrinsic rewards, such as parties and other entertainments. Nevertheless, the earlier sacrifices and resultant spiritual commitments helped Zoar commune last eighty-one years.

CONCLUSIONS

Both Albert Einstein and Bertolt Brecht (Chapters 4 and 6) had caring mothers, providing the relatedness that can foster lifelong intrinsic commitments. Einstein's mother had him navigate a busy Munich city street alone at age four. She issued many more challenges, such as learning the violin at age six, challenges that he surmounted. These successes no doubt increased his sense of efficacy so important in creating and preserving intrinsic commitments.[23]

Brecht's mother was convinced of his special aptitudes. She nurtured his self-confidence and encouraged his early writing, especially poetry. The warm relationship probably provided many extrinsic rewards that eventually led to internalization and the resulting intrinsic commitment to writing. He completed over 1,000 poems as well as more than fifty plays.[288]

Brecht's political commitment was less to Communism and more to helping the downtrodden. Thus he criticized the Communist East German government when they used tanks to put down the 1953 workers' uprising. That Brecht was not fully committed to Communism was also shown by his earlier reluctance to tarry in Moscow in 1941 while fleeing the Nazis.

Earlier, during the 1930s, Brecht apparently began to have some misgivings about Communism, and he created his stiff "teaching-learning" plays. A similar temporary escalation of commitment occurs when professionals such as teachers or social workers suffer from burnout, feeling unable to care about their clients. Given the prospect of having to give up their chosen work, there is often a surge in commitment, which may overcome the unpleasant burnout feelings.[289]

Like so many intrinsically motivated people, both Einstein and Brecht hated external controls by authorities who often use extrinsic rewards. Einstein fought with tutors and teachers. He could never enjoy music until he was free to teach himself. Einstein was expelled from his high school and Brecht nearly expelled.

Both Einstein and Brecht had further difficulties that they

successfully surmounted. In Einstein's case, his failing the entrance exam for the Swiss Institute of Technology plus his later inability to find a job must have energized his commitment to uncover new understandings in physics. The earlier successful overcoming of challenges set by his mother gave him an important start in developing an intrinsic commitment to overcome later challenges as a young adult. He was indeed a "Valiant Swabian."

Einstein's three decades of stubborn refusal to accept probabilistic quantum mechanics has many precedents among scientists. According to Max Planck, a new scientific truth is often not convincing to opponents. Scientists with strong intrinsic commitment to their own theories often have to die off before a new paradigm can become widely accepted.[30,p28]

The German Jews in Chapter 5 had a strong intrinsic commitment to Germany before Hitler took power in 1933. They had experienced prior decades of economic freedom, with resultant feelings of efficacy. Not surprisingly, most were reluctant to give up their commitment to being German citizens.

The German farmers in Chapter 3 came to Nebraska possessing a stubborn commitment to farming, common among the Swabian peasants. Not known is how much of the commitment had been fostered by the warmth of parents who initially extended extrinsic rewards and how much came from the children subsequently being put to work with some autonomy and positive feedback. These farmers, like Einstein, had the motivation to overcome many severe challenges, including even settling in a new country.

The intermittent nature of rewards in farming, with some prosperous years and some not, probably also played a role in preserving their continued commitment to farming in America. Farm laborers work for pay, and once the pay stops, they leave. Intrinsic commitments can be resistant to change because of the bonding of negative and positive results. Farmers can often feel that next year will be better.

The story of Immanuel Weiss (Chapter 7) shows how hard it is to kill the positive feelings experienced by intrinsically-motivated farmers. They find the freedom and the sense of efficacy wonderful. As one farmer explained to me, "no one can complain if I make a mistake."

Decline of Intrinsic Commitments

Some Nebraska farmers and their children later found the nature of farm work changing and becoming more like factory work. They had less autonomy. Pressures came from farms having to get bigger. The emphasis on crops over animals also fostered a relative loss in self-direction. It was no wonder that their intrinsic motivation came into question.

If the pressured farmers could not find an alternative satisfactory commitment, such as a focus on the family or perhaps helping their children escape farming, depression could result. Men sometimes medicate depression with alcohol.[30, 80]

German Jews (Chapter 5) also found their intrinsic motivations eroding. The increasingly restrictive Nazi laws decreased autonomy and the related sense of efficacy. Many Jewish refugees told me how proud and committed they used to be as German citizens. But after Hitler's taking power in 1933, Jews found full participation as German citizens more difficult. Nevertheless, intrinsic motivation once created can be hard to break. It took the unprecedented violence of Crystal Night to make most Jews question their commitment to remaining German. However, some, like Julius Guggenheim, were so committed that they returned from America to live in Germany after the war's end.

Germans have a stronger sense of duty to the nation than do Americans.[290] Some of the former Hitler Youths in Chapter 8 are examples. However, Immanuel Weiss (Chapter 7) did not have an excessive sense of such duty to Germany perhaps because he grew up in Romania. He even refused his promotion to corporal in the German army.

Fortunately, strong religious faith helped Weiss endure and not become depressed when he felt that he had lost his family. It is sometimes said that everyone is a believer in a foxhole, but Weiss already had a strong faith before his army experience. When I talked to him after he had come to America, he still had that strong intrinsic commitment.

Heinz Biesdorf gradually rejected a commitment to the Nazis as a youth, so it was not surprising that he became cynical about fighting for his country, nor did he have a religious faith that could carry him through the war like Weiss. He simply wanted to survive.

From Extrinsic to Intrinsic Commitment

Eva Bromke (Chapter 8) grew up with an initial extrinsic commitment to religion like most Swabian children. In Germany, religion is a subject at school, but Bromke found there was a difference between grades being given (extrinsic rewards) and learning religion from the priest who was a good person, supplying the relatedness necessary for an intrinsic acceptance of Catholicism.

When Bromke had questions about doctrines like the Immaculate Conception, she increased her efforts to maintain commitment, like many others suffering doubts or burnout. She "turned many pages" in her search for answers.

Fortunately, Bromke had the freedom to explore and found her explorations gratifying. She came to her own conclusions, eventually being able to retain her commitment to Catholicism, but in her own idiosyncratic fashion. Like some other Catholics, she rejected the rules of the institutional church but continued her spiritual commitment.[291] Like Rudy Geiger, also in Chapter 8, she satisficed (settled for something satisfactory) rather than optimized.[34]

Geiger had little autonomy as a child. He always had to work at jobs. He had no freedom to develop an intrinsic commitment to farming. Instead, he picked up a strong extrinsic commitment, to money.

In Geiger's business in America, he reacted like a typical Type A personality. He felt pressure to make more and more money, and he responded by trying to preserve complete control. He planned every step of the work, its timing, and the assignments of all the men. He sought perfect efficiency. This led to health problems, fortunately short of the heart attacks suffered by many Type As.[292]

Geiger had the freedom to explore ways to release the pressure he felt to become a millionaire. He tried giving up some extrinsic rewards by letting one worker take over part of the business. Eventually he found a better solution by converting some of his extrinsic motivation into intrinsic motivation. Whenever there is some choice involved in selecting among extrinsic rewards, with autonomy and efficacy involved, some intrinsic motivation can develop. He eventually chose to focus on the process, the positive feelings he got from good workmanship. His need for extrinsic rewards then lessened. He could then charge a fair price rather than trying to earn as much money as possible.

Clearly, commitments are mutable. Extrinsic commitments can develop into intrinsic commitments. In Geiger's case, his new approach to work was a change in focus, from external to internal. Studies show that people similarly suffering burnout sometimes try to solve their problem by making sudden and drastic changes, such as a banker joining a nonprofit organization. Fortunately, Geiger did not have to reject his line of work and give up the skills he had perfected.

Like many Swabians, Heinz and Ellen Biesdorf (Chapter 8) developed a commitment to the norms of conscientiousness fostered by the punitive church courts during the Protestant Reformation. Even after the courts ended, frugality remained a societal norm in Swabia. For most people, it became an automatic habit. Such norms are reinforced in the home and schools. Like most norms, if the behaviors are over-learned by repetition, they can become automatic.[23] Then, if there is some choice or autonomy involved, such habits may evolve into internally satisfying, intrinsic motivations.

The Biesdorfs freely chose to continue their commitment to

frugality in America despite the American norms of not caring about saving. The Biesdorfs did this even after becoming wealthy. When I last talked to them, they admitted they were finding it hard to spend down their savings. Intrinsic motivation is difficult to curtail.

Intrinsic Motivation in Swabia

There are three major cultural emphases by which Swabia differs from the rest of Germany—individualism, spirituality, and conscientiousness.[5] These have probably helped form some of the strong intrinsic motivation we have seen here.

Probably the most significant basis for the intrinsic motivation found in people like Einstein, Brecht, and other Swabians, is the widespread feeling of individual autonomy. This became a common norm in Swabia over the many centuries during which the economy was based solely on independent farming. Working on one's own land also fostered a sense of efficacy, the second important basis for intrinsic motivation.[23]

A typical Swabian joke has a peasant receiving a visit from the local pastor who noted how he had built up his farm into a veritable Garden of Eden, saying, "You see how God's grace has been granted you." The peasant responds, "Yes, you should have seen this place before, when God was the only one working it."

Even though Swabia was historically agricultural, many eminent inventors and entrepreneurs have appeared in recent times: Daimler and Maybach (internal combustion engine), Benz (motorcar), Bosch (automobile ignitions), Diesel and Wankel (innovative engines), as well as Walther and Mauser (firearms). Such discoveries and inventions usually need stubborn single-mindedness.

A study of 710 inventors showed that their main trait was perseverance. The study concluded that working onlyeight hours a day, five days a week would never make a person an inventor.[293] Scientists also commonly possess such single-mindedness. A study of sixty-four scientists showed that they had in common an absorption in their work. "They have worked long hours for many years, frequently with no vacations to speak of because they would rather be doing their work than anything else." Such persistent dedication is not sufficient to foster scientific eminence, but it seems to be necessary.[294,p51]

Today's Swabian part of Germany has become a land of high tech industries. It leads the rest of Europe in patents per capita and it is second behind Silicon Valley in world sales of business software and computer services. It also has the greatest number of research institutes in Europe.[295] Perhaps the Swabian cultural norm of autonomy and the resultant intrinsic motivation have helped foster these achievements. Work in modern organizations is no longer routine, but complex. Research shows the old-fashioned approach to motivating workers with extrinsic rewards is no longer effective.[296] Today, where employees increasingly have to work smarter, intrinsic motivation is essential.

The widespread sense of personal autonomy in Swabia has probably also helped foster the intrinsic motivation of its many accomplished writers and thinkers such as Friedrich Schiller, Friedrich Hölderlin, Ludwig Uhland, Justinus Kerner, Eduard Mörike, and Wilhelm Hauff. Many specialized in poetry, not exactly a monetarily rewarding form of writing. The modern writer, Hermann Hesse, will be most familiar to Americans. Eminent Swabian philosophers include Friedrich Schelling and, G.W.F. Hegel.

In an experiment with seventy-two young creative writers, some were told of the extrinsic rewards of writing such as getting into graduate school. Others got intrinsic motivation from hearing that people can write even in the absence of external goals or pressures. This experiment revealed that even a suggestion of extrinsic rewards can hinder creative absorption and result in less-accomplished writing.[108]

In addition to autonomy, another important aspect of Swabian culture is its strong spirituality. The more than two centuries of church courts used extrinsic motivation in the form of punishments to foster religious observance. However, the later popularity of

Pietism meant that people met in small groups away from religious authorities. This allowed individuals to develop personal forms of spiritual commitment.

Research shows that many Swabians see life as having meaning because of the existence of God. In addition, sects are more common in Swabia than in the rest of Germany, probably also demonstrating continuing intrinsic commitment to spirituality by some. This is part of the long tradition in Swabia of resistance against the authoritarian strictures of the Catholic and Lutheran churches with their reliance on extrinsic rewards and punishments. [5]

Perhaps this intrinsic commitment to spirituality among some Swabians has occasionally transformed into the intrinsic motivations of the eminent writers and inventors mentioned above. Hermann Hesse's life showed how an early religious background can develop into intrinsic motivation for writing and philosophizing.

A third important part of Swabian culture came from the punitive religious courts and their fostering of conscientiousness—austerity, frugality, tidiness, prudence, scrupulousness, honesty, and hard work. The norms created subsequently had the potential to develop into intrinsic motivation, depending on whether people could choose to observe the norms or not.

Immanuel Weiss's story (Chapter 7) shows how consistent application of extrinsic rewards can foster automatization of behavioral norms that may develop into intrinsic motivation. He had much physical punishment as a child. German parents felt that sparing the rod spoils the child. The widespread use of punishment apparently fostered the strong commitment to honesty that Weiss noted among the Germans in contrast to the Romanians. If there were dishonest dealings, one's reputation would suffer. This was a form of extrinsic control.

However, a component of intrinsic motivation in the commitment to honesty could develop if there was later choice involved—if at some point a person could freely decide whether to be honest or dishonest. Successful creation of intrinsic motivation means the external requirements are seen as compatible with one's values and are freely chosen. Then children would not feel controlled. They will feel they are acting autonomously. The process comes to have intrinsic value and is personally meaningful and enduring. The Biesdorfs' frugality entailed this internalization of extrinsically rewarded behaviors. In contrast, honesty created solely by extrinsic rewards can weaken.[297]

One account of a European approach to childrearing suggests how conversion of extrinsic motivation to intrinsic motivation may be possible. In France, parents and teachers are initially quite strict, using rewards and punishments. There is an emphasis on obedience. Then freedoms are gradually granted during adolescence to allow the formation of intrinsic motivation for important values.[298]

In America, Amish parents are similarly strict at first, but then allow a time of complete freedom in adolescence. Youths will often run wild, smoking, drinking, and engaging in other previously prohibited behaviors. However, because of the autonomy granted then, most decide to return to the Amish community, having developed an intrinsic commitment to Amish values.[299]

A common approach in America is just the opposite—much permissiveness and freedom for young children and then an attempt to clamp down on sex and alcohol during adolescence. Unfortunately, by that time many young people have become used to, and expect, continued great freedom. The abuse of alcohol among high school and college youths is just one result.

There has been some public discussion about how to foster "character" or values among young people. Internalization can happen if parents or teachers show warmth and are nurturing toward young people, thus creating a feeling of connectedness. Research shows that teens exposed to cold, controlling maternal care are more likely to develop materialistic extrinsic orientations, while better nurtured children come to value the intrinsic goals of personal growth, relationships, and community.

As has been stressed above, it is important there be some choice, some autonomy and not complete control, regarding the values to be internalized. Autonomy allows successful internalization so individuals come to accept that formerly imposed extrinsic goals are their own. Unfortunately, parents and teachers may be seduced by the apparent ease of controlling simply with rewards and punishments. Developing character cannot be achieved if only extrinsic motivations are stressed.[297, 300, 301]

Intrinsic Commitments in America?

Intrinsic commitments seem to have become less common in modern America. Until the twentieth century, the Protestant Work Ethic postulated work as an intrinsic end in itself.[289] Many people felt a specific calling, perhaps to minister to a congregation, to doctor the sick, or to be a teacher or social worker for those who needed help. They agreed with Einstein: "Work is the only thing that gives substance to life."[10,p457] The traditional calling was for service without major extrinsic rewards like good pay, and it involved sacrifices such as self-discipline, perseverance, and other forms of conscientiousness. This did not always come easy, as ego strength has limits.[302]

The work ethic and associated asceticism have had a long history in Germany, which became known as the "nation of work." However, even there the work ethic has faded as many modern Germans have turned their backs on religion. Germany seems like America, and America like Germany, in losing this strong, religious-based belief in the value of work as an end in itself.[303]

A major hindrance for intrinsic motivation today is that many important extrinsic rewards are available. Daniel Bell pointed out in *The Cultural Contradictions of Capitalism* that hard work in capitalism produced plenty, and that ended the Spartan life of someone committed to the work ethic. As we noted above, intrinsically motivated students getting money for doing homework can lose interest if the money is removed. The money focuses one's attention externally, not on internal satisfactions. Extrinsic rewards can sap intrinsic commitments.

Even traditional intrinsic commitments like love and the family seem to have diminished somewhat in America. Temporary relationships and divorce are more common. These are sometimes due to the feeling that too many sacrifices are being demanded, interfering with personal happiness.

Today, intrinsic commitments are often seen as old-fashioned. In the most popular social science book of all time, *The Lonely Crowd*, the authors described people who were "inner-directed," with a gyroscope that enabled them to follow values learned early in life. These people were confident, but rigid. These seem to be people who are stubbornly committed to intrinsically-motivated values. Another social type described was the "tradition-directed," people who followed traditions just because they were traditions.

Both of these orientations have been mostly replaced by the "other-directed." Americans have become more flexible, more willing to adjust to others in order to gain their approval. As society has moved from independent farmers, craftsmen, and entrepreneurs and become comprised of large organizations, other-direction has become more necessary. People have become more committed to extrinsic rewards such as money, allowing them to move flexibly within and between organizations. People with the rigidity of intrinsic commitments can be misfits in an organizational society.[8]

There has also been an increased emphasis in America on self-identity—such as self-actualization, self-esteem, and fame.[304] People will say, "I have to be me" or "What's in it for me?" People are reluctant to incur the sacrifices that can solidify commitments that transcend the self. Wayne Dyer in his book, *Your Erroneous Zones*, told people to outlaw self-denial unless it is absolutely necessary, and it rarely is.

Even commitments to a career have seen a greater emphasis on rewarding results rather than the intrinsic commitment to the process. As former football coach Woody Hayes of Ohio State put it, "We'd rather have an immoral win than a moral victory."[23,p248]

The aggrandizement of the self seems to promise lasting happiness. However, there is a major problem. The self eventually has to die. There is no transcendent commitment. As Goethe noted in his play about Faust, "God help us—art is long, and our life so short."

Many individuals in today's consumer society also focus on various extrinsic rewards that they mistakenly view as permanently fulfilling. People often ignore the hedonic treadmill. Getting an extrinsic reward will usually result in adaptation. One soon has to seek more rewards to keep the same level of happiness. Television and other forms of entertainment, for example, are often only temporarily satisfying but often yield a feeling of emptiness only remedied by still more entertainment.[26, 305]

One version of the Faust story stresses the temporary nature of extrinsic rewards. Dr. Faust, a Swabian, confidently agreed that once he received complete joy and knowledge he would let the Devil have his soul. Faust knew the Devil would never be successful in providing lasting satisfaction.

The pursuit of happiness seems desirable and understandable, but if all action is for self-interest, the greater good for society may be neglected. A result of unbridled individualism has been termed the "tragedy of the commons." This refers to the ancient tradition of the village commons where villagers could let their animals graze. However, clever people often put more and more of their animals on the commons, reaping personal benefits (extrinsic rewards). However, this naturally led others to follow suit, destroying the pasture for everyone.

Actions that are rational for the individual can be contrary to the interests of society. Only an intrinsic commitment to preserving the commons, the greater entity, can prevent destructive rational self-interest.[306] Fortunately, some people have today embraced an intrinsic commitment to preserving our environment before it is too late.

An observation from Goethe confirms the importance of such

intrinsic commitments: "There are but two roads that lead to an important goal and to the doing of great things: strength and perseverance. Strength is the lot of but a few privileged men; but austere perseverance, harsh and continuous, may be employed by the smallest of us and rarely fails of its purpose, for its silent power grows irresistibly greater with time."[307] Such is the potential power of intrinsic commitment.

Some of the world's greatest triumphs and achievements have been the result of *Eigensinn* or single-mindedness. The Greeks long ago recognized this. Hesiod in the sixth century BC wrote, "Before the gates of excellence the high gods have placed sweat; long is the road thereto and rough and steep."[308,p76]

Of course, not everyone may want to be so dedicated to an intrinsic commitment. However, most people are free to cultivate such commitments in moderation, free to develop intrinsic motivation for satisfactions not dependent on the external rewards that can be taken away or that can dissipate in potency. Many have the potential to revel in the associated sense of efficacy and the accompanying "flow" in which time passes most pleasantly. If not in work, then intrinsic commitments can be developed for interpersonal relationships or for hobbies. Even in small ways, intrinsic satisfactions are available to all of us.

SOURCES

1. Katz, D. and K. Braly, *Racial stereotypes of one hundred college students.* The Journal of Abnormal and Social Psychology, 1933. **28**(3): p. 280.
2. Peabody, D., *National characteristics.* 1985, Cambridge [Cambridgeshire] ; New York: Cambridge University Press ix, 256 p.
3. Goethe, J.W.v., *Götz von Berlichingen.* 1882. pp. 181.
4. Wikipedia. Available from: http://en.wikipedia.org/wiki/Swabian_salute.
5. Wieland, G.F., *Celtic Germans: The rise and fall of Ann Arbor's Swabians.* 2014, Ann Arbor: G.F. Wieland. 395 p.
6. Helmreich, W.B., *The things they say behind your back: Stereotypes and the myths behind them.* 1982: Transaction Publishers.
7. Schmitt, D.P., et al., *The geographic distribution of Big Five personality traits patterns and profiles of human self-description across 56 nations.* Journal of Cross-Cultural Psychology, 2007. **38**(2): p. 173-212.
8. Riesman, D., *The lonely crowd; a study of the changing American character.* 1956, Garden City, N. Y.,: Doubleday. 359 p.
9. Brehm, S.S. and J.W. Brehm, *Psychological reactance : a theory of freedom and control.* 1981, New York: Academic Press. xiii, 432 p.
10. Einstein, A., *The ultimate quotable Einstein.* 2011: Princeton University Press.

11. Csikszentmihalyi, M., *Beyond boredom and anxiety*. 2000, San Francisco: Jossey-Bass Publishers. 231 p.
12. Vallerand, R.J., *From motivation to passion: In search of the motivational processes involved in a meaningful life*. Canadian Psychology/Psychologie canadienne, 2012. **53**(1): p. 42-52.
13. Deci, E.L., R. Koestner, and R.M. Ryan, *A meta-analytic review of experiments examining the effects of extrinsic rewards on intrinsic motivation*. Psychological Bulletin, 1999. **125**(6): p. 627-668.
14. Ochse, R., *Before the gates of excellence : the determinants of creative genius*. 1990, Cambridge ; New York: Cambridge University Press. ix, 300 p.
15. Shils, E.A. and M. Janowitz, *Cohesion and disintegration in the Wehrmacht in World War II*. Public Opinion Quarterly, 1948. **12**(2): p. 280-315.
16. Baumeister, R.F. and J. Tierney, *Willpower : rediscovering the greatest human strength*. 2011, New York: Penguin Press. 291 p.
17. Ruderman, A.J., *Dietary restraint: a theoretical and empirical review*. Psychological Bulletin, 1986. **99**(2): p. 247-262.
18. Mata, J., et al., *Motivational "spill-over" during weight control: Increased self-determination and exercise intrinsic motivation predict eating self-regulation*. Health Psychology, 2009. **28**(6): p. 709-716.
19. Williams, G.C., et al., *Testing a self-determination theory process model for promoting glycemic control through diabetes self-management*. Health Psychology, 2004. **23**(1): p. 58-66.
20. Curry, S., E.H. Wagner, and L.C. Grothaus, *Intrinsic and extrinsic motivation for smoking cessation*. Journal of Consulting and Clinical Psychology, 1990. **58**(3): p. 310-316.
21. Williams, G.C., et al., *Testing a self-determination theory intervention for motivating tobacco cessation: supporting autonomy and competence in a clinical trial*. Health Psychology, 2006. **25**(1): p. 91-101.

22. Deci, E.L., *Effects of externally mediated rewards on intrinsic motivation.* Journal of Personality and Social Psychology, 1971. **18**(1): p. 105.
23. Klinger, E., *Meaning & void : inner experience and the incentives in people's lives.* 1977, Minneapolis: University of Minnesota Press. xiv, 412 p.
24. Amabile, T.M., B.A. Hennessey, and B.S. Grossman, *Social influences on creativity: the effects of contracted-for reward.* Journal of Personality and Social Psychology, 1986. **50**(1): p. 14-23.
25. Frankl, V.E., *Man's search for meaning; an introduction to logotherapy.* A newly rev. and enl. ed. 1962, [New York, N. Y.]: Simon and Schuster. 151 p.
26. Richins, M.L., *Media, materialism, and human happiness.* Advances in Consumer Research, 1987. **14**(1): p. 352-356.
27. Kasser, T. and R.M. Ryan, *A dark side of the American dream: correlates of financial success as a central life aspiration.* Journal of Personality and Social Psychology, 1993. **65**(2): p. 410-422.
28. Kasser, T. and R.M. Ryan, *Further examining the American dream: Differential correlates of intrinsic and extrinsic goals.* Personality and Social Psychology Bulletin, 1996(22): p. 80-87.
29. Sheldon, K.M. and T. Kasser, *Pursuing personal goals: Skills enable progress, but not all progress is beneficial.* Personality and Social Psychology Bulletin, 1998. **24**(12): p. 1319-1331.
30. Brickman, P., R.M. Sorrentino, and C.B. Wortman, *Commitment, conflict, and caring.* 1987, Englewood Cliffs, N.J.: Prentice-Hall. xiv, 317 p.
31. Deci, E.L., *The psychology of self-determination.* 1980: Free Press. x,240.
32. Ryan, R.M. and E.L. Deci, *Self-determination theory and the facilitation of intrinsic motivation, social development, and well-being.* American Psychologist, 2000. **55**(1): p. 68-78.
33. Baard, P.P., Deci, E. L., Ryan, R. M., *Intrinsic need satisfaction: A motivational basis of performance and well-being in work settings.* 1998: Fordham Univrsity.
34. March, J.G. and H.A. Simon, *Organizations.* 1958, New York: Wiley. 262 p.

35. Aronson, E. and J. Mills, *The effect of severity of initiation on liking for a group.* The Journal of Abnormal and Social Psychology, 1959. **59**(2): p. 177.
36. Festinger, L., *When prophecy fails.* 1956, Minneapolis,: University of Minnesota Press. vii, 256 p.
37. Kanter, R.M., *Commitment and community; communes and utopias in sociological perspective.* 1972, Cambridge, Mass.,: Harvard University Press. x, 303 p.
38. Zimbardo, P.G. and A.B. Cross, *Stanford prison experiment.* 1971: Stanford University.
39. Milgram, S., *Behavioral study of obedience.* The Journal of Abnormal and Social Psychology, 1963. **67**(4): p. 371-378.
40. Wallace, P.A.W., *Conrad Weiser, 1696-1760 : friend of colonist and Mohawk.* 1945, Philadelphia,: University of Pennsylvania press;. xiv, 648 p.
41. Graeff, A.D., *Conrad Weiser: Pennsyvania peacemaker.* 1945, Allentown, PA: Schlechter's. 395p.
42. Canipe, M., *Swabians on the Carolina frontier.* 2013: Mack Edward Canipe. 260p.
43. Leder, L.H., *Robert Livingston, 1654-1728: and the politics of colonial New York.* 1961: University of North Carolina Press.
44. Kapp, F., *Geschichte der deutschen einwanderung in Amerika.* 1867, New York,: E. Steiger. vii, 410 p.
45. Wertenbaker, T.J., *The founding of American civilization : the Middle Colonies.* 1963, New York: Cooper Square Publishers. xiii, 364 p.
46. Durnbaugh, D.F., *Radical Pietist Involvement in Early German Emigration to Pennsylvania.* Yearbook of German-American Studies, 1994. **29**: p. 29-48.
47. Arndt, K.J.R., *Harmony on the Connoquenessing, 1803-1815 : George Rapp's first American Harmony : a documentary history* 1980, Worcester, Mass.: Harmony Society Press. xliv, 1021 p.
48. Arndt, K.J.R., *George Rapp's successors and material heirs.* 1971, Rutherford [N.J.]: Fairleigh Dickinson University Press. 445 p.
49. Tomney, J.J., *Divine eonomy: George Rapp, the Harmony Society and Jacksonian democracy.* 2014, Liberty University: Lynchburg, VA. p. 125.

50. Arndt, K.J.R., *George Rapp's Harmony Society, 1785-1847*. Rev. ed. 1972, Rutherford [N.J.]: Fairleigh Dickinson University Press. 713 p.
51. Arndt, K.J.R., *George Rapp's Harmony Society, 1785-1847*. 1965, Philadelphia,: University of Pennsylvania Press. 682 p.
52. Wilson, W.E., *The angel and the serpent; the story of New Harmony*. 1964, Bloomington,: Indiana University Press. xiv, 242 p.
53. Rapp, G., *Thoughts on the destiny of man, particularly with reference to the presnt times*. 1924, New Harmony, Indiana: Harmony Society. 96.
54. Büttner, J.G., *Briefe aus und über Nordamerika: oder, Beiträge zu einer richtigen kenntniss der Vereinigten Staaten und ihrer bewohner, besonders der deutschen bevölkerung, in kirchlicher, sittlicher, socialer und politischer hinsicht, und zur beantwortung der frage über auswanderung, nebst nachrichten über klima und krankheiten in diesen staaten*. 1845: Arnold.
55. Arndt, K.J.R., *Harmony on the Wabash in transition, 1824-1826 : transitions to George Rapp's Divine Economy on the Ohio, and Robert Owen's New Moral World at New Harmony on the Wabash : a documentary history*. 1982, Worcester, Mass.: Harmony Society Press. xl, 876 p.
56. Holloway, M., *Heavens on earth: Utopian communities in America, 1680-1880*. 1966: Courier Corporation.
57. Knortz, K., *Amerikanische Lebensbilder*. 1884, Zürich,: Verlags-Magazin (J. Schabelitz). 2 p. L., 208 p.
58. Duss, J.S., *The Harmonists: a personal history*. 1970, Ambridge, PA: Harmonie Associates. 425.
59. Nordhoff, C., *The Communistic societies of the United States : from personal visit and observation, including detailed accounts of the Economists, Zoarites, Shakers, the Amana, Oneida, Bethel, Aurora, Icarian and other existing societies, their religious creeds, social practices, numbers, industries, and present condition*. 1875, New York [N.Y.]: Harper. viii, 439 p.
60. Sutton, R.P., *Communal utopias and the American experience: religious communities, 1732-2000*. 2003: Greenwood Publishing Group.

61. Fogarty, R.S., *All things new: American communes and utopian movements, 1860-1914*. 2003: Lexington Books.
62. Randall, E.O., *History of the Zoar society, from its commencement to its conclusion; a sociological study in communism*. 3d ed. 1904, Columbus, O.,: Press of F. J. Heer. 2 p. l., 105 p.
63. Gorisek, S., *From a to Zoar*, Ohio Magazine. 1989. p. 27-33, 74.
64. Joy, R.K., Ruth; Hueftle, Bonnie; Boerkircher, Bev; Keller, Dianne; McClatchey, Diann; Englebrecht, Emma; Wolf, Jo; and Smallfoot, Sharon, ed. *1776-1976: Eustis, progressive bicentennial community*. 1976, Loup Valley Queen Printers: Callaway, NE.
65. Dick, E.N., *Conquering the Great American Desert : Nebraska*. 1975, [Lincoln]: Nebraska State Historical Society. xiii, 456 p.
66. Hine, R.V., *The American West; an interpretive history*. 1973, Boston,: Little, Brown. x, 371 p.
67. Bartlett, R.A., *The new country: a social history of the American frontier, 1776-1890*. 1974: Oxford University Press London.
68. Smith, H.N., *Virgin land; the American West as symbol and myth*. 1950.
69. Blank, W.e.a., *Depressive Erkrankungen bei Schwaben und Heimatvertriebenen*. Der Nervenarzt, 1981. **52**: p. 153-162.
70. Fawcett, E. and T. Thomas, *The American condition*. 1982: HarperCollins Publishers.
71. Coughenour, C.M. and L.E. Swanson, *Rewards, values, and satisfaction with farm work*. Rural Sociology 1988. **55**(4): p. 442-459.
72. Perman, E., *The effect of ethyl alcohol on the secretion from the adrenal medulla in man*. Acta Physiologica Scandinavica, 1958. **44**(3-4): p. 241-247.
73. Cronk, C.E. and P.D. Sarvela, *Alcohol, tobacco, and other drug use among rural/small town and urban youth: a secondary analysis of the monitoring the future data set*. American Journal of Public Health, 1997. **87**(5): p. 760-764.
74. McClelland, D.C., et al., *The drinking man: Alcohol and human motivation*. 1972, New York: Freepress.

75. Marlatt, G.A., *Alcohol, stress, and cognitive control*, in *Stress and anxiety*, I.G. Sarason, Editor. 1976, John Wiley: New York. p. 271-296.
76. Stivers, R., *Historical meanings of Irish-American drinking*, in *The American experience with alcohol: contrasting cultural prspectives*, L.A. Bennett, Editor. 1985, Plenum: New York. p. 109-129.
77. Greeley, A.M., W.C. McCready, and G. Theisen, *Ethnic drinking subcultures*. 1980, New York: Praeger. iii, 138 p.
78. Scoufis, P. and M. Walker, *Heavy drinking and the need for power.* Journal of Studies on Alcohol and Drugs, 1982. **43**(09): p. 1010-1019.
79. Hingson, R., T. Mangione, and J. Barrett, *Job characteristics and drinking practices in the Boston metropolitan area.* Journal of Studies on Alcohol and Drugs, 1981. **42**(09): p. 725-38.
80. Neff, J.A. and B.A. Husaini, *Life events, drinking patterns and depressive symptomatology; the stress-buffering role of alcohol consumption.* Journal of Studies on Alcohol and Drugs, 1982. **43**(03): p. 301-18.
81. Einstein, A. and A.P. French, *Einstein : a centenary volume*. 1979, Cambridge, Mass.: Harvard University Press. xx, 332 p.
82. Vallentin, A., *The drama of Albert Einstein*. 1954, Garden City: Doubleday.
83. Fölsing, A., *Albert Einstein : a biography*. 1997, New York: Viking. xiii, 882 p.
84. Veithans, H., *Arbeiten zum historischen Atlas von Südwestdeutschland.* Vol. V. 1970, Stuttgart: W. Kohlhammer.
85. Hoffmann, B. and H. Dukas, *Albert Einstein, creator and rebel*. 1972, New York: Viking Press. xv, 272 p.
86. Brockman, J., *My Einstein : essays by twenty-four of the world's leading thinkers on the man, his work, and his legacy.* 1st ed. 2006, New York: Pantheon Books. xvi, 261 p.
87. Uther, H.-J., *Handbuch zu den Kinder-und Hausmärchen" der Brüder Grimm*. 2008: Walter de Gruyter Berlin/New York.

88. Stone, A.D., *Einstein and the quantum : the quest of the valiant Swabian*. 2013, Princeton etc.: Princeton University Press. 332 S.
89. Pyenson, L. and A.J. Meadows, *The young Einstein : the advent of relativity*. 1985, Bristol ; Boston: A. Hilger. xiv, 255 p.
90. Calaprice, A., *The expanded quotable Einstein*. 2000: Princeton University Press Princeton, NJ.
91. Hoffmann, B. and H. Dukas, *Albert Einstein : creator and rebel*. 1973, New York: New American Library. xv, 272 p.
92. Clark, R.W., *Einstein:: the life and times*. 1984: Avon.
93. Pais, A., *Subtle is the lord-- : the science and the life of Albert Einstein*. 1982, Oxford [Oxfordshire] ; New York: Oxford University Press. xvi, 552 p.
94. Schilpp, P.A., *Albert Einstein: Autobiographical Notes*, in *La Salle, Illinois: Open Court Publishing Company. First published in PA Schilpp ed., Albert Einstein: Philosopher-Scientist in The Library of Living Philosophers*. 1979.
95. Dukas, H. and B. Hoffmann, *Albert Einstein, the human side: new glimpses from his archives*. 1979, Princeton: Princeton Univ. Press. 167.
96. Jammer, M., *Einstein and religion : physics and theology*. 1999, Princeton, NJ: Princeton University Press. 279 p.
97. Einstein, A., *Cosmic religion : with other opinions and aphorisms*. 1931, New York: Covici-Friede. 109 p.
98. Neffe, J., *Einstein : a biography*. 2007, New York: Farrar, Straus, and Giroux. x, 461 p.
99. Smolin, L., *The other Einstein*. New York Review of Books, 2007. **54**(10(June 14)).
100. Bucky, P.A., A. Einstein, and A.G. Weakland, *The private Albert Einstein*. 1992, Kansas City: Andrews and McMeel. xii, 171 p.
101. Frank, P., *Einstein: his life and times*. 2002: Da Capo Press.
102. Willems, E., *A way of life and death: three centuries of Prussian-German militarism: an anthropological approach*. 1986, Nashville: Vanderbilt University Press. 226 p.
103. Cohen, I.B., *Einstein and Newton*, in *Einstein: a centenary volume*, A.P. French, Editor. 1979, Harvard University Press: Cambridge. p. 40-49.

104. Fox, K.C. and A. Keck, *Einstein : A to Z*. 2004, Hoboken, N.J.: J. Wiley. x, 310 p.
105. Highfield, R. and P. Carter, *The private lives of Albert Einstein*. 1993, London ; Boston: Faber and Faber. xii, 355 p.
106. Einstein, A., A. Calaprice, and A. Einstein, *The new quotable Einstein*. Enlarged commemorative ed. 2005, Princeton, NJ: Princeton University Press.
107. Einstein, A., *Ideas and opinions. Based on Mein Weltbild*. 1954, New York: Crown Publishers. 377 p.
108. Briggs, J., *Fire in the crucible : the self-creation of creativity and genius*. 1990, Los Angeles: J.P. Tarcher. 382 p.
109. Isaacson, W., *Steve Jobs*. 2011, London: Little, Brown. 630 p.
110. Overbye, D., *Einstein in love : a scientific romance*. 2000, New York: Viking. xv, 416 p.
111. Tocqueville, A.d., *Democracy in America. 2 vols*. New York: Vintage, 1945.
112. Simon, P.-H., *From pacifism to the bomb*, in *Einstein*, L. de Broglie, Editor. 1979, Peeples Press. p. 151-171.
113. Freud, S., *Why war?(1932)*. Standard Editon of the Psychological Work of Sigmund Freud, 1964. **22**: p. 203-15.
114. Isaacson, W., *Einstein : his life and universe*. 2007, New York: Simon & Schuster. xxii, 675 p.
115. Deuel, W.R., *People under Hitler*. 1942, New York,: Harcourt. viii, 392 p.
116. Snow, C.P., *Einstein*, in *Einstein: the first hundred years*, M. Goldsmith, Editor. 1980, Pergamon: Oxford. p. 3-18.
117. Straus, E.G., *Memoir*, in *Einstein: a centenary volume*, A.P. French, Editor. 1979, Harvard University Press: Cambridge. p. 31-32.
118. Frank, P. and S. Kusaka, *Einstein : his life and times*. 1953, New York: Knopf. 298, xii p.
119. Aczel, A.D., *Entanglement : the greatest mystery in physics*. 2002, New York: Four Walls Eight Windows. xviii, 284 p.
120. Adler, L., *Israelitische Religionsgemeinschaft of Wurttemberg its development and changes*. The Leo Baeck Institute Yearbook, 1960. **5**(1): p. 287-298.
121. Dicker, H., *Creativity, Holocaust, reconstruction: Jewish life in Wuerttemberg, past and present*. 1984, Brooklyn: Sepher Hermon Press. 234.

122. Tänzer, A., *Die Geschichter der Juden in Württemberg.* 1937, Frankfurt: J. Kaufmann. xv, 190.
123. Jeggle, U., *Judendörfer in Württemberg.* 1969, Tübingen,: Tübinger Vereinigung f. Volkskunde e. V. 361 p.
124. Weinryb, B.D., *The German Jewish immigrants to America (a critical evaluation)*, in *Jews from Germany in the United States*, E.E. Hirschler, Editor. 1955, Farrar, Straus: New York. p. 101-127.
125. Barkai, A., *German-Jewish migrations in the nineteenth century, 1830-1910.* Leo Baeck Institute Yearbook, 1985: p. 301-318.
126. Glanz, R., *The German Jewish mass emigration: 1820-1880.* American Jewish Archives, 1970. **22**(52): p. 1-4.
127. Schwarz, S., *Die Juden in Bayern im Wandel der Zeiten.* 1980, München: G. Olzog. 368 p.
128. Wiesemann, F., *Die Jüdischen Gemeinden in Bayern 1918-1945: Geschichte u. Zerstörung:[Veröffentlichung im Rahmen d. Projekts"Widerstand u. Verfolgung in Bayern 1933-1945."* 1979: Oldenbourg.
129. Kussy, S., *Reminiscensce of Jewish life in Newark, N.J.* Yivo Annual of Jewish Social Science. **6**.
130. Stephenson, C., *The process of community: class, culture, and ethnicity in nineteenth-century Newark*, in *New Jersey's ethnic heritage*, P.A. Stellhorn, Editor. 1978, New Jersey Historical Commission: Trenton. p. 94-132.
131. Lowenstein, S.M., *The rural community and the urbanization of German Jewry.* Central European History, 1980. **13**(03): p. 218-236.
132. Glanz, R., *The"Bayer"and the"Pollack"in America.* Jewish Social Studies, 1955: p. 27-42.
133. Waxman, C.I., *America's Jews in transition.* 1983, Philadelphia: Temple University Press. xxv, 272 p.
134. Cahnman, W.J., *Village and small-town Jews in Germany: a typological study.* The Leo Baeck Institute Yearbook, 1974. **19**(1): p. 107-130.
135. Picard, J., *Childhood in the village: fragment of an autobiography.* The Leo Baeck Institute Yearbook, 1959. **4**(1): p. 273-293.
136. Schwab, H., *Jewish rural communities in Germany.* 1956, London: Cooper Book Company.

137. Stern, B., *Meine Jugenderinnerungen an eine württembergische Kleinstadt und ihre jüdische Gemeinde: mit einer Chronik der Juden in Niederstetten und Hohenlohe vom Mittelalter bis zum Ende des Zweiten Weltkriegs.* Vol. 4. 1968: Kohlhammer.
138. Frank, F., *Kindheit in Horb.* Schäbische Heimat, 1977. **28**((1)Jan-March): p. 42-51.
139. Gay, P., *Freud, Jews, and other Germans : master and victims in modernist culture.* 1978, New York: Oxford University Press. xx, 289 p.
140. Kaplan, M.A., *Between dignity and despair : Jewish life in Nazi Germany.* 1998, New York: Oxford University Press. xii, 290 p.
141. Laqueur, W., *The German Youth Movement and the 'Jewish Question'A Preliminary Survey.* The Leo Baeck Institute Yearbook, 1961. **6**(1): p. 193-205.
142. Rhodes, J.M. and A. Hitler, *The Hitler movement : a modern millenarian revolution.* 1980, Stanford, Calif.: Hoover Institution Press. 253 p.
143. Mosse, G.L. and M.A. Ledeen, *Nazism : a historical and comparative analysis of National Socialism.* 1978, New Brunswick, N.J.: Transaction Books. 134 p.
144. Peukert, D., *Inside Nazi Germany : conformity, opposition and racism in everyday life.* 1987, New Haven: Yale University Press. 288 p.
145. Fritzsche, P., *Life and death in the Third Reich.* 2008, Cambridge, Mass.: Belknap Press of Harvard University Press. viii, 368 p.
146. Fritz, A., *Die Geschichte und Entwicklung der Juden in Buttenhausen.* 1938, Hochschule Hohenheim: Hohenheim. p. 62.
147. Strauss, W., *Signs of life, Jews from Wuerttemberg : reports for the period after 1933 in letters and descriptions.* 1982, New York: Ktav. 389 p.
148. Prittie, T., *Germans against Hitler.* 1964, London,: Hutchinson. 291 p.
149. Wirth, G., *Stuttgarts Beitrag zur Kunst der Gegenwart*, in *Stuttgarter Kunst im 20. Jahrhundert*, H. Heissenbuettel, Editor. 1979, Deutsche Verlags-Anstalt: Stuttgart. p. 72-135.

150. Rombach, O., *Zwischen Murrhardt und New York.* An Rems Murr, 1974. **1**: p. 43-49.
151. Read, A. and D. Fisher, *Kristallnacht: unleashing the Holocaust.* 1989: Michael Joseph. 294.
152. Stokes, L.D., *The German people and the destruction of the European Jews.* Central European History, 1973. **6**(02): p. 167-191.
153. Mack, J., *Nationalism and the self.* Psychohistory Review, 1983: p. 47-69.
154. Bankier, D., *The Germans and the final solution : public opinion under Nazism.* 1992, Oxford, UK ; Cambridge, Mass., USA: B. Blackwell. vii, 206 p.
155. Anonymous. *A letter by a firefighter.* Eyewitness accounts and reminiscences [Website] 1938 [cited 2015 May 23, 2015]; Available from: http://motlc.wiesenthal.com/site/pp.asp?c=gvKVLcMVIuG&b=394831.
156. Sauer, P., *Württemberg in der Zeit des Nationalsozialismus.* 1975, Ulm: Süddeutsche Verlagsgesellschaft. 519 p.
157. Sauer, P., *Die Schichksale der jüdischen Bürger Baden-Württembergs während der nationalsozialistischen Verfolgungszeit 1933-1945.* 1969, Stuttgart: W. Kohlhammer.
158. Schadt, J., *Verfolgung und Widerstand.* Das Dritte Reich in Baden und Württemberg, hg. v. Otto Borst, Stuttgart, 1988: p. 96-120.
159. Prinz, A., *The role of the Gestapo in obstructing and promoting Jewish emigration.* Yad Vashem Studies, 1958. **2**: p. 205-218.
160. Landau, R.S., *The Nazi Holocaust.* 1994, Chicago: I.R. Dee. xiv, 356 p.
161. Gellately, R., *The Gestapo and German society: political denunciation in the Gestapo case files.* The Journal of Modern History, 1988: p. 654-694.
162. Gellately, R., *The Gestapo and German society: enforcing racial policy, 1933-1945.* 1982, New York: Oxford University Press.
163. Marrus, M.R. and R.O. Paxton, *The Nazis and the Jews in occupied Western Europe, 1940-1944.* The Journal of Modern History, 1982: p. 687-714.
164. Arendt, H., *Eichmann in Jerusalem; a report on the banality of evil.* 1963, New York,: Viking Press. 275 p.

165. Kelman, H.C., *Crimes of obedience : toward a social psychology of authority and responsibility.* 1989, New Haven: Yale University Press. xiii, 382 p.
166. Kershaw, I., *Popular opinion and political dissent in the Third Reich, Bavaria 1933-1945.* 1983, Oxford, Oxfordshire Clarendon Press xii, 425 p.
167. Koch, E., *Deemed suspect: A wartime blunder.* 1985: James Lorimer & Company.
168. Naegele, P., *Forward* in *Health and healing*, E. Cumming, Editor. 1970, Jossey-Bass: San Francisco.
169. Mills, T.M. and F.E. Jones, *Naegele, Kaspar, D.-1923-1965- In Memoriam.* 1965, American Sociological Review. p. 579-581.
170. Rombach, O., *Zwischen Murrhardt und New York: Über den Maler Reinhold Nägele.* An Rems Murr, 1974. **1**: p. 43-49.
171. Heißenbüttel, H., *Stuttgarter Kunst im 20. Jahrhundert. Malerei, Plastik, Architektur,* Stuttgart, 1979.
172. Eberle, J., *Die Reise nach Amerika:.* 1949, Stuttgart: Der Turmhaus Druckerei.
173. Levine, L.G., *Dancing Girl.* 1992, Frankfurt am Main: Haag and Herchen. 87 p.
174. Dobkowski, M., *4th Reich-German-Jewish religious life in America today.* Judaism, 1978. **27**(1): p. 80-95.
175. Cherfas, L., et al., *The framing of atrocities: Documenting and exploring wide variation in aversion to Germans and German-related activities among Holocaust survivors.* Peace and Conflict: Journal of Peace Psychology, 2006. **12**(1): p. 65-80.
176. Einstein, K., *People power and how to choose the powerful people.* Executives' Club News, 1979. **55**(5(October 26)): p. 1-8.
177. Shukert, E.B. and B.S. Scibetta, *The war brides of World War II.* 1988, Novato, CA: Presidio Press. 302 p.
178. Messinger, H., *Schwäbisch.* 2004, München: Polyglott.
179. Esslin, M., *Brecht: the man and his work.* 1971, Garden City, N.Y.: Anchor Books. xix, 379 p.
180. Hayman, R., *Brecht: A biography.* 1983, Oxford University Press: New York. 423 p.
181. Hill, C., *Bertolt Brecht.* 1975, Boston: Twayne Publishers. 208 p.

182. Brecht, B., et al., *Poems [of] Bertolt Brecht: 1913-1956*. 1976, New York: Methuen.
183. Münsterer, H.O., *The young Brecht*. 1992, London: Libris. 195 p.
184. Parker, S., *Bertolt Brecht: a literary life*. 2014, London: Bloomsbury Methuen Drama.
185. Murphy, G.R., *Brecht and the Bible: A Study of Religious Nihilism and Human Weakness in Brecht's Drama of Mortality and the City*. 1980: University of North Carolina Press.
186. Grimm, R., *Luther's Language in the Mouth of Brecht: A Parabolic Survey with Some Examples, Detours, and Suggestions*. Michigan Germanic Studies Ann Arbor, Mich., 1984. **10**(1-2): p. 159-204.
187. Murphy, G.R., *Brecht and the Bible : a study of religious nihilism and human weakness in Brecht's drama of mortality and the city*. 1980, Chapel Hill: University of North Carolina Press. 104 p.
188. Münsterer, H.O., *Recollections of Brecht in 1919 in Augsburg*, in *Brecht as they knew him*, H. Witt, Editor. 1974, International Publishers: New York. p. 23-31.
189. Willett, J., *The Poet beneath the Skin*. Brecht heute—Brecht Today: Jahrbuch der Internationalen Brecht-Gesellschaft, 1972. **2**: p. 88-104.
190. Völker, K., *Brecht, a biography*. 1978, New York: Seabury. 412 p.
191. Needle, J. and P. Thomson, *Brecht*. 1981, Chicago: University of Chicago Press. 235.
192. Morley, M., *Brecht : a study*. 1977, Totowa, N.J.: Rowman and Littlefield. 135 p.
193. Guttmann, W. and P. Meehan, *The great inflation, Germany 1919-23*. 1975, Westmead, Farnborough, Hants, England: Saxon House.
194. Friedrich, O., *Before the deluge; a portrait of Berlin in the 1920's*. 1972, New York,: Harper & Row. xi, 418 p.
195. Ewen, F., *Bertolt Brecht; his life, his art, and his times*. 1967, New York: Citadel Press. 573 p.
196. Merkl, P.H., *The making of a stormtrooper*. 1980, Princeton, N.J.: Princeton University Press. xix, 328 p.

197. Scheff, T.J., *Bloody revenge : emotions, nationalism, and war*. 1994, Boulder: Westview Press. xi, 162 p.
198. Esslin, M., *Brecht, a choice of evils : a critical study of the man, his work, and his opinions*. 1980, London: Eyre Methuen. xvi, 315 p.
199. Schoeps, K.-H., *Bertolt Brecht*. 1977, New York: Frederick Ungar.
200. Hamilton, R.F., *Who voted for Hitler?* 1982, Princeton, N.J.: Princeton University Press. xv, 664 p.
201. Mosse, G.L., *Nazi culture: Intellectual, cultural and social life in the Third Reich*. 1966: Univ of Wisconsin Press.
202. Rees, G., *The great slump: capitalism in crisis, 1929-1933*. [1st U.S. ed. 1971, New York,: Harper & Row. 310 p.
203. Dicks, H.V., *Licensed mass murder; a socio-psychological study of some SS killers*. Columbus Centre series. 1972, New York,: Basic Books. xiii, 283 p.
204. Weinstein, F., *The dynamics of Nazism : leadership, ideology, and the holocaust*. 1980, New York: Academic Press. xviii, 168 p.
205. De Jonge, A., *The Weimar chronicle : prelude to Hitler*. 1978, New York: Paddington Press : distributed by Grosset & Dunlap. 256 p.
206. Whiting, C., *The home front: Germany*. Vol. 32. 1982, Chicago: Time Life.
207. Lyon, J.K., *Bertolt Brecht in America*. 1980, Princeton, N.J.: Princeton University Press. 408 p. .
208. Bentley, E., *The Brecht commentaries, 1943-1980*. 1981, New York: Grove Press. 320 p.
209. Hook, S., *Out of step: An unquiet life in the 20th century*. 1987: Harper & Row New York.
210. Grund, F.J., *The Americans, in their moral, social, and political relations*. Vol. 2. 1837, London,: Longman, Rees, Orme, Brown, Green & Longman. 2 v.
211. Dundes, A., *Life is like a chicken coop ladder : a portrait of German culture through folklore*. 1984, New York: Columbia University Press. xi, 174 p.
212. Brecht, W., *Unser leben in Augsburg, damals : Erinnerungen*. 1985, Frankfurt am Main: Insel. 367 p.
213. Cook, B., *Brecht in exile*. 1983, New York: Holt, Rinehart and Winston. xiii, 237 p.

214. Bilderback, W.a.C., Scott T., *A stage of learning: The Brecht Company.* Theatre, 1985. **16**(3(Summer/Fall)): p. 27-35.
215. Eddershaw, M., *Acting methods: Brecht and Stanislavsky*, in *Brecht in perspective*, G. Bartram, Editor. 1982, Longman: New York. p. 128-44.
216. Brecht, B., *Parables for the theatre: Two plays: The good woman of Setzuan and The caucasian chalk circle.* 1948, Minneapolis: University of Minnesota Press.
217. Lewin, M., *The impact of Kurt Lewin's life on the place of social issues in his work.* Journal of Social Issues, 1992. **48**(2): p. 15-29.
218. Johnson, P., *Intellectuals.* 1988, New York: Harper & Row. x, 385 p.
219. Lewin, K., *Some social-psychological differences between the United States and Germany.* Journal of Personality, 1936. **4**(4): p. 265-293.
220. Whitaker, P., *Brecht's poetry : a critical study.* 1985, Oxford: Clarendon Press. 284 p.
221. Kebir, S., *Ein akzeptabler Mann? : Brecht und die Frauen.* 1989, Koln: Pahl-Rugenstein. 194 p.
222. Lenya, L., *August 28, 1928*, in *Bertolt Brecht: The Threepenny Opera*, E. Bentley, Editor. 1960, Grove Press: New York.
223. Berlau, R. and H. Bunge, *Living for Brecht: the memoirs of Ruth Berlau.* 1987: Fromm International.
224. Ritchie, G.F., *Der Dichter und die Frau.* 1989, Bonn: Bouvier. vii, 394p.
225. Fuegi, J., *Brecht and company : sex, politics, and the making of the modern drama.* 1994, New York: Grove Press. 732 p.
226. Johnson, N., *That book should be burned: the contested afterlife of Bertolt Brecht.* Journal of Postgraduate Research, 2006. **5**: p. 8-19.
227. Feuchtwanger, L., *Bertolt Brecht presented to the British*, in *Brech as they knew him*, H. Witt, Editor., International Publishers: New York. p. 17-22.
228. Gray, R.D., *Brecht the dramatist.* 1976, Cambridge [Eng.] ; New York: Cambridge University Press. 232 p.
229. Weiss, I. and G.F. Wieland, *Bessarabian knight: a peasant caught between the red star and the swastika: Immanuel*

Weiss's true story. 1991, Lincoln, NE: American Historical Society of Germans from Russia. 149 p.
230. Wagner, I., *Zur Geschichte der Deutschen in Bessarabien.* 1958, Stuttgart: Heimatmuseum der Deutschen in [ie aus] Bessarabien.
231. Becker, J., *Bessarabien und sein Deutschtum.* 1966, Bieigheim/Wurttemberg: Krug.
232. Shafir, M., *Romania, politics, economics, and society : political stagnation and simulated change*. 1985, Boulder, Colo.: L. Rienner Publishers. xvi, 232 p.
233. Baumann, O., *The reasons for the emigration of our forefathers.* Journal of the American Historical Society of Germans from Russia, 1989. **12**(4(Winter)): p. 44-50.
234. Height, J.S., *Homesteaders on the steppe, cultural history of the Evangelical-Lutheran colonies in the region of Odessa, 1804-1945*. 1975, Bismarck, N.D.: North Dakota Historical Society of Germans from Russia. xii, 431 p.
235. Knell, V., *The life and work of the Bessarabien German women in the old homeland.* Heritage Review, 1987. **17**((4)December): p. 18-21.
236. Treichel, D.E., *Heimatbuch der Gemeinde Kulm*. n.d., Frankfurt/Main: H.G. Gachet. 284 p.
237. Seton-Watson, H., *Eastern Europe between the wars, 1918-1941*. 1945, Cambridge [Eng.]: The University Press. 442 p.
238. Steinert, M.G., *Hitler's war and the Germans ; public mood and attitude during the Second World War*. 1977, Athens: Ohio University Press. x, 387 p.
239. Kater, M.H., *Hitler in a Social Context.* Central European History, 1981. **14**(03): p. 243-272.
240. Lowie, R.H., *Toward understanding Germany*. 1954, Chicago: University of Chicago Press. ix, 396 p.
241. Eberhardt, E., *The Bessarabian German Dialect in Medicine Hat, Alberta*. 1973, Thesis (Ph. D.)--University of Alberta.
242. Read, A. and D. Fisher, *The deadly embrace : Hitler, Stalin, and the Nazi-Soviet Pact, 1939-1941*. 1988, London: Michael Joseph. xvi, 687 p.
243. Daniel, H., *The ordeal of the captive nations*. 1958, Garden City, NY: Doubleday.
244. Conquest, R., *The harvest of sorrow : Soviet collectivization and the terror-famine*. 1986, London: Hutchinson. 412 p.

245. Schechtman, J.B., *European population transfers, 1939-1945*. 1946: Oxford University Press New York.
246. Ziemke, E.F. and Time-Life Books., *The Soviet juggernaut*. 1980, Alexandria, Va: Time-Life Books. 208 p.
247. Simons, G. and Time-Life Books., *Victory in Europe*. World War II. 1982, Alexandria, Va.: Time-Life Books. 208 p.
248. Ziemke, E.F., *Stalingrad to Berlin: the German defeat in the east*. 1968, Washington,: Office of the Chief of Military History [for sale by the Supt. of Docs. xiv, 549 p.
249. Time-Life, *The SS (Third Reich)*. 1989, Richmond VA: Time-Life Books. 186 p.
250. De Zayas, A.M., *Nemesis at Potsdam : the Anglo-Americans and the expulsion of the Germans : background, execution, consequences*. 1977, London ; Boston: Routledge & K. Paul. 268 p.
251. Solzhenitsyn, A.I., *The Gulag Archipelago, 1918-1956; an experiment in literary investigation*. [1st ed. 1974, New York,: Harper & Row. 3 v.
252. Seydewitz, M., *Civil life in wartime Germany, the story of the home front*. 1945, New York,: The Viking press. viii, 448 p.
253. Collingham, E.M., *The taste of war : World War Two and the battle for food*. 2011, London: Allen Lane. xv, 634 p.
254. Marrus, M.R., *The unwanted : european refugees in the twentieth century*. 1985, New York: Oxford University Press. xii, 414 p.
255. Germany (West). Bundesministerium für Vertriebene Flüchtlinge und Kriegsgeschädigte. and T. Schieder, *Dokumentation der Vertreibung der Deutschen aus Ost-Mitteleuropa*. 1953, [Bonn,. 5 v. in 8.
256. Schieder, T. and Germany (West) Bundesministerium für Vertriebene Flüchtlinge und Kriegsgeschädigte., *Die Vertreibung der deutschen Bevölkerung aus der Tschechoslowakei*. Dokumentation der Vertreibung der Deutschen aus Ost-Mitteleuropa. 1957, [Bonn,. 2 v.
257. Botting, D., *From the ruins of the Reich : Germany, 1945-1949*. 1985, New York: Crown. 341 p.
258. Bailey, R.H. and Time-Life Books., *Prisoners of war*. World War II. 1981, Alexandria, Va.: Time-Life Books. 208 p.

259. Botting, D. and T.-L. Books, *The aftermath: Europe*. 1983, Alexandria, VA; Morristown, NJ: Time-Life Books.
260. Ryder, A.J., *Twentieth-century Germany: from Bismarck to Brandt*. 1973, London,: Macmillan. xx, 656, 16 p.
261. Grosser, A., *Germany in our time. Translated by Paul Stephenson*. 1971, Praeger: New York.
262. Douglas, R.M., *Orderly and humane: The expulsion of the Germans after the Second World War*. 2012: Yale University Press.
263. Sorge, M.K., *The other price of Hitler's war : German military and civilian losses resulting from World War II*. 1986, New York: Greenwood Press. xx, 175 p.
264. Mayer, K.U., *German survivors of World War II*. Social Structures and Human Lives., 1988: p. 229-246.
265. Bernard, W.S., *Immigration: history of US policy*, in *Harvard encyclopedia of American ethnic groups*. 1980, Belknap: Cambridge. p. 486-495.
266. Nerger-Focke, K., *Die deutsche Amerikaauswanderung nach 1945: Rahmenbedingungen und Verlaufsformen*. Vol. 14. 1995: H.-D. Heinz.
267. Schneider, K., *Schule und Erziehung*, in *Das Dritte Reich in Baden und Württemberg*, O. Borst, Editor. 1988, Konrad Theiss: Stuttgart. p. 121-136.
268. Koch, H.W., *The Hitler Youth: origins and development 1922-45*. 1975, London,: Macdonald and Jane's. xi, 340 p.
269. Reichsjugendführung, *Pimpf im Deinst: ein Handbuch für das Deutsche Jungvolk in der HJ*. 1938, Potsdam: L. Voggenreiter.
270. Schoenbaum, D., *Hitler's social revolution; class and status in Nazi Germany, 1933-1939*. 1966, Garden City, N.Y.,: Doubleday. xxiii, 336 p.
271. Bormann, P. and D. Fritzsche, *The Schirmacher Oasis, Queen Maud Land, East Antarctica, and its surroundings*. Ergänzungsheft 289, Petermanns geographische Mitteilungen. 1995, Gotha: J. Perthes. 448 p.
272. Kater, M.H., *Hitler youth*. 2009, Cambridge: Harvard University Press.
273. Rumpf, H., *The Bombing of Germany*. New York: Holt, Rinehart and Winston, 1962: 256 p.

274. MacIsaac, D. and U.S.S.B. Survey, *The United States Strategic Bombing Survey: a collection of the 31 most important reports printed in 10 volumes*. 1976: Garland.
275. Hewitt, K., *Place annihilation: area bombing and the fate of urban places*. Annals of the Association of American Geographers, 1983. **73**(2): p. 257-284.
276. United States Strategic Bombing Survey., *The effects of strategic bombing on German morale*. 1947, Washington,. 2 v.
277. Sherry, M., *The Slide to Total Air War*. The New Republic, 1981. **16**.
278. Beck, E.R., *The Allied bombing of Germany, 1942-1945, and the German response: Dilemmas of judgment*. German Studies Review, 1982: p. 325-337.
279. Peterson, E.N., *The American occupation of Germany : retreat to victory*. 1977, Detroit: Wayne State University Press. 376 p.
280. Davis, F.M., *Come as a conqueror; the United States Army's occupation of Germany, 1945-1949*. 1967, New York,: Macmillan. xvi, 271 p.
281. Zink, H., *The United States in Germany, 1944-1955*. 1957, Princeton, N.J.,: Van Nostrand. 374 p.
282. Solzhenitsyn, A.I., *The first circle*. 1968, New York,: Harper & Row. xiii, 580 p.
283. Friedländer, S., *Nazi Germany and the jews: The years of extermination: 1939-1945*. 2007, HarperCollins: New York. 870 p.
284. Main, J., *Lessons from the supershoppers*. Money, 1980(June): p. 66-68, 70.
285. Kantor, L.S.a.K.L., Alden Manchester, Victor Oliveira, *Estimating and addressing America's food losses*. Food Review, 1997. **20**(1(January-April)): p. 2-12.
286. Anonymous, *Consumer classrooms by the millions: an interview with Heinz B. Biesdorf*. Human Ecology Forum, 1980(Fall): p. 13-15.
287. Zajonc, R.B., *Attitudinal effects of mere exposure*. Journal of Personality and Social Psychology, 1968. **9**(2p2): p. 1-27.
288. Brecht, B., et al., *Poems, 1913-1956*. 1979, New York: Methuen. xxvii, 627 p.

289. Baumeister, R.F., *Meanings of life*. 1991, New York: Guilford Press. xii, 426 p.
290. McClelland, D.C., et al., *Obligations to self and society in the United States and Germany*. The Journal of Abnormal and Social Psychology, 1958. **56**(2): p. 245-255.
291. Bromley, D.G., *Falling from the faith: Causes and consequences of religious apostasy*. Vol. 95. 1988: Sage Publications, Inc.
292. Friedman, M. and R.H. Rosenman, *Type A behavior and your heart*. [1st ed. 1974, New York,: Knopf. 276 p.
293. Rossman, J., *The psychology of the inventor : a study of the patentee*. 1931, Washington, D.C.: Inventors Pub. Co. 252 p.
294. Roe, A., *A psychologist examines sixty-four eminent scientists*, in *Creativity: selected readings*, P.E. Vernon, Editor. 1970, Penguin: Harmondsworth. p. 23-51.
295. Trübenbach, M., *Die Superlative der Region Stuttgart und Baden-Württemberg*. n.d., Ostfildern: Matthias Trübenbach.
296. Likert, R., *New patterns of management*. 1961, New York: McGraw-Hill. 279 p.
297. Deci, E.L. and R.M. Ryan, *Facilitating optimal motivation and psychological well-being across life's domains*. Canadian Psychology/Psychologie canadienne, 2008. **49**(1): p. 14-23.
298. Carroll, R., *Cultural misunderstandings : the French-American experience*. 1988, Chicago: University of Chicago Press. xiii, 147 p.
299. Shachtman, T., *Rumspringa : to be or not to be Amish*. 2006, New York: North Point Press. 286 p.
300. Kasser, T., et al., *The relations of maternal and social environments to late adolescents' materialistic and prosocial values*. Developmental Psychology, 1995. **31**(6): p. 907-914.
301. Brooks, D., *Love and merit*, in *New York Times*. 2015: New York. p. A2.
302. Baumeister, R.F., et al., *Ego depletion: is the active self a limited resource?* Journal of Personality and Social Psychology, 1998. **74**(5): p. 1252.
303. Campbell, J., *Joy in work, German work : the national debate, 1800-1945*. 1989, Princeton, N.J.: Princeton University Press. 431 p.

304. Bellah, R.N., *Habits of the heart : individualism and commitment in American life*. 1985, Berkeley: University of California Press. 355 p.
305. Kubey, R.W. and M. Csikszentmihalyi, *Television and the quality of life : how viewing shapes everyday experience*. Communication. 1990, Hillsdale, N.J.: L. Erlbaum Associates. xvii, 287 p.
306. Hardin, G., *The tragedy of the commons*. Science, 1968. **162**(December 13): p. 1243-1248.
307. Müller, F.v., *Goethe in seiner practische Wirksamkeit; eine Vorlesung in der Academie Gemeinnütziger Wissenschaften zu Erfurt am 12, September 1832*. 1832, Weimar: W. Hoffmann. 46 p.
308. Hamilton, E., *The lessons of the past*, in *Aventures of the mind*, R. Thruelson, Editor. 1960, Gollancz: London. p. 69-80.

INDEX

Aachen, 204
Achern, 144
action, by audience, 171, 172
action, Swabian, 71
Adler, Fritz, 110, 122, 132
Adler, Karl: to America, 131; attacked by Nazis, 108; collaboration with Gestapo, 119; cultural programs end, 118; fights for Germany, 107; fired from job, 108; friends of, 109; Gestapo letters to, 136; grilled by Gestapo, 120, 121; imprisoned, 118; musical career, 107, 108; as Nazi, 136; reports on emigrations, 119; visits Germany, 135
Adorno, Theodor, 165
Advanced Study, Institute for, 81, 91, 93
affairs (sexual): Brecht's 174; Einstein's, 85, 93. *See also* Bahnholzer, Paula; Berlau, Ruth; Hauptmann, Elizabeth; Steffin, Margarete; Weigel, Helene; Zoff, Marianne.
African Americans, 92
agreeableness, 2
Air Corps, German, 222
airplanes: 219, 222; fighter, 200, 228, 229, 231; Stuka, 197
Albany, New York, 22

alcohol: and adrenaline, 65; and alcoholics, 65; and boredom, 66; excessive, in Eustis, 58; hiding stills, 59; making, 53; restrictions on, 238; rural usage, 65; and sense of power, 66; and sex roles, 66; and toughness, 66; tradition of, 54; and youths, 66, 67
aliens: British internments, 132; British restrictions, 129; German, in U.S., 167, 168; Jews in America, 133, 134
Allegheny Mountains, 21, 23
Allies, as bullies, 185
All Quiet on the Western Front, 158
Alsace, 130
Alsace-Lorraine, 89
Amana communes, 31, 40, 41
Ambridge PA, 42
Amendment, First, 168
America: industrial power of, 222; Jews to, 124, 128, 130, 137; Naegele's reactions to, 133; and pacifism, 143
America/Germany, cultural differences, 171–74
Americanization, 56
Americans: agreeableness of, 2; criticism disliked by, 171; Einstein criticizes, 88; team with, 227
Amish, 42, 259
Ammer, K. L., 175
Anabaptists, 42
anger, Brecht's, 145
animals, slaughtering of, 190
Anne, Queen of England, 14, 15
annexations, of German territory, 200
Antarctica, 181
anxieties, 7
apprenticeships, 225

Arabs (Palestinian), 88
Arbeitsdienst, 222
Argentina, 166
Aristotle, 172
armies: American, 202; British, 204; German mass conscription, 185; Romanian, 185, 186, 188, 189
arrogance, 201, 236; Einstein's, 70
Aryanization of businesses, 114
Aryans (Gentiles), 106
asceticism, Weiser's, 21
assassinations, character, 176
assertiveness, Prussian, 80
assimilation, 99, 182; r, 184
Atlantic Charter, 124
Atonement, Day of (*Yom Kippur*), 128
Augsburg: Brecht's birthplace, 144; Brecht a theater critic, 165; Brecht in military, 148; Brecht leaves, 154; Eva Bromke in, 228–30, 232–37
Aurora commune, 42
austerity, fiscal, 158
Austria, 80, 127, 169, 227
authorities: and Biesdorf, 224; Brecht handles, 145, 168, 169; Bromke questions, 239; and Einstein, 70, 72, 74, 79; German respect for, 120, 230; and Laughton, 167; in Lutheran/Catholic Churches, 258; Pietists free from, 258; use extrinsic rewards, 251
authority, Einstein as, 84
autobahns, 184
autonomy: by Amish, 259; Biesdorf seeks, 224; for children, 8, 9; to choose farming, 65; and Einstein, 70, 71; to explore, 254; in farming, 47, 252; habits and intrinsic motivation, 255; of Harmonists, 38; for intrinsic motivation, 7, 249; from intrinsic motivation, 10; lack of, 254; and norms, 258; in Swabia, 256, 257; Geiger's, 255; by writers/inventors, 257

aviators, 228
Baal, 151, 152, 177
Bach, Johann Sebastian, 136
Baden, 101, 144
Bahnholzer, Paula, 148, 149, 151
Baisingen, 101
Baker, R. L., 38, 39
Balkans, 146
Balts, 192
Banat, 184
bank runs, 158
Barbarossa, King, 71
bar mitzvah, 115, 116
basement shelters, 229
battlefields, World War I, 89
Bäumeler, Joseph. *See* Bimeler, Joseph
Bavaria, 226
Bavaria: King of, 149; Republic, 149; Soviet, 150
Beaver Falls PA, 40
beer, 54
behaviorism, 155
Beissel, Conrad, 19
Belgium: 90, 205; Queen of, 72
beliefs, religious, 7
Bengel, Johann Albrecht, 32
Bentley, Eric, 163, 170, 172
Berlau, Ruth, 163, 174, 175
Berlin: bombing of, 223; Brecht ill in, 145; Brecht to, 154; class-conflicts in, 80; Communist revolutions in, 150; East, 164, 176, 177; Einstein's apartment, 84, 85; Einstein to, 79, 89; orders fight to end, 233; Polish

INDEX

Jews in, 87; rebellions in, 148; Soviet/British sectors of, 208; University of, 81, 89
Berliners, to Weiss farm, 206
Bessarabia: ethnic groups in, 181, 185, 186; invaded by Soviets, 186, 187; lost farm in, 214; Russian, 184
Bethlehem PA, 20
Bible, 58, 145, 147, 171
Bible, The, 146
Biesdorf, Ellen, 245, 246
Biesdorf, Heinz, 217–25, 245–48, 254
Biesdorfs, 11, 255, 256, 259
Bimeler, Joseph, 42, 44
Black Forest, 144, 241
black markets, 210, 235
Blitzkrieg invasions, 193
Bloch, Bella, 102
Blue Mountain, 17
Blue Ridge Mountains, 21
boarding houses, 102
Bohemian Corporal, 127
Böhme, Jacob, 31, 43
Bohr, Neils, 94, 95
Bolsheviks, 149
bombings: of Berlin, 202; of Britain, 193; carpet and strategic, 227, 229; explosive and incendiary, 230; fleeing from, 206; Germans as victims of, 228; of oil/factories, 222; radio alerts for, 128; of Stuttgart, Berlin, 223; working around, 225
bombs, atomic, 78, 91
bookkeeping, 61, 66
books: burned, 112, 113, 160; on war, 90
border controls, 128
Bosch (firm) 120
bosons, 94

boycotts, 112
Braddock, Gen. Edward, 23
Brandenburg, East, 200
brandy, 59
Brecht, Bertolt: 143-180; accepted by public, 179; birth, 144; childhood, 144; close to mother, 144; criticism by, 143; as leftist, 143; moralistic, 143; parents' backgrounds, 144; relatedness for, 251; stubborn in exile, 1; tinkering by, 11, 176
Brecht, Walter, 148
breeding, of cattle, 63
brides, war, 236–37
Britain, resists Hitler, 193
British Broadcasting Corporation, 222
Broadway, 133, 160, 163, 166, 167
Bromke, Chester, 237–39
Bromke, Eva, 11, 228–41, 254
Bronx, 134
brooding, Swabian, 70
Brownshirts. *See* Storm Troopers
Brünning, Chancellor Hermann, 158
Buchau, 70, 72
Bulgaria, 193
Bulgarians, 181, 188, 189
bums, 237
bunkers, air raid, 128
bureaucratization, 230
Burlington and Missouri Railroad, 49
burnout, 251, 254, 255
businesses: Einsteins', 75; Jewish, 117, 130
Buttenhausen, 106, 107, 141
California Institute of Technology, 90
Calley, Lieutenant William, 125
callings, 96, 173, 260
Calvinism, 176
Calvin, John, 81

Canada, 133
Canasatego, 22
Cannstadt, 116, 128
capitalism: in America, 164; Brecht's criticism of, 176; in Germany, 155; poisons humans, 156; and urban problems, 155
caring, after the revolution, 174
Carolinas, 14, 15
Carol, King of Romania, 187
Casualties: German, Brecht's mother on, 222; German and Soviet, 193, 194
Catholics: and Bromke, 239-41; favor Hitler less, 234; teachings questioned, 239, 240; and Weiser, 20
cattle, 63
cattlemen, 48
Caucasian Chalk Circle The, 166, 170
causes, in quantum mechanics, 94
Cayuga Indians, 16
celibacy: at Harmony, 33, 36, 37; and Hutterites, 41; at Zoar, 44
Central Intelligence Agency, 165
Central Pacific Railroad, 51
cents, two and a half, 242
certainty, 83, 154
chain migrations, 48, 49
challenges: 10; and Einstein, 93; on farms, 252
change: Brecht's commitment to, 180; in Brecht's plays, 169; in Communism, 155; continual, 177; deferred, in *Spartacus*, 149; in farming, 62
chaos, postwar, 206
Chaplin, Charlie, 85
charisma, Hitler's, 185
chauvinism, male, 61
Chicago, 155
children: American, 212, 259; Brecht's, 149, 151, 163; and farming, 57, 64, 65, 67, 68, 182, 252; habits of, 12; hours of work, 225; illegitimate, 93; and intrinsic motivation, 259; during Nazi era, 217; Nazis exclude, 115; and parental relatedness, 252; punishment of, 183; repatriation of, 191; transports, 122; Weiss, 190, 202, 206, 212, 213
Christianity, Brecht's, 145
Christians, 204
Christ, Jesus, 143, 162, 210
Christmas, 43, 54, 105, 115, 130, 135-136, 195, 198
church courts, 255, 257
churches: attendance at, 60; and Crystal Night, 116; in Eustis, 54; and euthanasia, 242; and extrinsic motivation, 5, 12; voting in, 62
Church, Lutheran. *See* Lutheran Church
Church World Service, 212
citizens, election of, 179
citizenship: and Jews, 113; and Weiss family, 192
Civilian Conservation Corps, 219
classes: rigid, 220; social, 79
Cleveland, 43
Clinton, Governor, 22
closed endings, 172
clothing, 81
clowns, 92
Cold War, 168, 212
collaboration, 118, 174–76
Columbian County, Ohio, 32
commercialism of Hollywood, 164
Commie, Brecht as, 170, 179

INDEX

commitments: escalation of, 251; intrinsic, 7; of German Jews, 253; obsessional, Einstein's, 84; obsessional, Jobs' and Picasso's, 84, and values, 3

Committee of Intellectual Cooperation, 90

commodity exchanges, 155

common people, 166

communes, 1, 29–45, 250

Communism: aims, 178; Brecht's commitment to, 251; Brecht not a member, 169, 178; concern for oppressed, 143; Romanian, 215; as scientific, 155

Communist, anti-, Brecht as, 171

Communists: to concentration camps, 107, 160; discipline of, 156; and FBI, 168; join Nazis, 109; not a threat, 92; street battles by, 159; violence in Munich, 150

community, emphasized by Nazis, 241

compasses, 73

competence. *See* efficacy

comradeship, 156, 159, 217, 220

concentration camps, 219, 233

conferences, colonist/Indian, 18

Congress, U.S., 32, 33

Conrad Weiser State Park, 17

conscientiousness: Biesdorf's, 255; and farming, 65; and Swabian Jews, 138, 139

consciousness raising, 164

Constance, Lake, 225

Constitution, U.S., 22

consulates, U.S., 122, 123

consumption, efficiency in, 245

contacts, attraction from, 250

Continental Congress, 22

contradictions, 171

controls, external. *See* motivation, extrinsic

convictions, Brecht's, 157

cooking, 246

cooperation: international, 90; manipulated, 155

cooperatives, writing, 165

Cornell University, 245

corporals (army), 195

corrective systems, Communist, 155

Council Fire, Onondaga, 22

court-martials, 197

crashes, stock market, 158

creativity, 7

credit, buying on, 247

Crimea, 199

criticisms: American dislike of, 88; by Brecht, 150, 165, 168; value for Brecht, 177

Crystal Night: and Adler, 136, 137; and bar mitzvah, 116; ends Jewish commitment to Germany, 123, 253; and Hans Walz, 120; in Stuttgart, 117; triggers emigration, 122

Cuba, 131, 132

culinary theater, 170

culture conflicts, German-American, 171, 236

Culture, Minister of Science, Education and Popular, 218

Culture, Reich Chamber of, 113

currency reforms, 211

Czechoslovakia, 127, 160, 200

Dachau (concentration camp), 116, 117, 127, 233

dangers, Swabians avoid, 223

Danish (farmers), 56

Danube River, 187, 188, 192

dark energy, 96
Darwin, Charles, 11
Darwinism, Social, 224
decorum, in Germany, 235
decrees, rule by, 153, 160
Defense Law, for Jews, 113
Delaware Indians: and Iroquois, 26; on Ohio River, 23; scalpings by, 23, 24, 25; Zinzendorf and, 20
demonstrations (political), enjoying, 153
denazification, 234
Denmark, 161
dependence, Brecht's, 145, 174
deportations, 124
depression (psychiatric): and alcohol, 66; Brecht's, 163; and extrinsic motivation, 7; faith counteracts, 254; in farming, 253; Maric's, 84; and suicides, 125; in Swabia, 62; Weiss', 204
Depression, Great, 158
deserting army, 188, 189
Detroit, 220
devil's advocates, 177
dialectics: Brecht's, 177; Marxian, 155
dialectic theater, 177-78
dictionaries, 2
didactic plays. *See* teaching-learning plays
dieting, 5, 6
differences, German-America, 238
disarmament, German, 154
Displaced Persons Acts, 212
divorces, 84
Dniester River, 186
Dno, 197
dogmas, 240
dogmatism, fatherly, 146

doubts, 254
downtrodden, and Brecht, 251
drafts, military, 148, 194, 223
drawer, writing for the, 161
Dresden, 202
Drums in the Night, 151
drunkenness, of farmers, 65
Dukas, Helen, 85, 86, 93
dumb workers, 243
Dunkirk, 132
Duquesne University, 245
Duss, John and Susie, 41, 42
Dust Bowl, 57
duties, 218, 230, 253
Dyson, Freeman, 227
Easter, 183
Easter Bunnies, 54
East Germany. *See* Germany, East
East Prussia, 198, 199, 200
Eberle, Josef, 134
Ebert, Friedrich, 149
education, Montessori, 8
efficacy: from choice, 255; Einstein's, from challenges, 8, 252; in farmers, 47; in farming, 59, 67, 252; for intrinsic motivation, 7, 249, 251; from intrinsic satisfaction, 10; in Swabia, 256
efficiency, 224, 225, 245–48
egalitarianism, 79, 80, 165
egoism, 152
Einöd, 49
Einstein, Albert, 69–97; avoiding extrinsic motivation, 69; decades of stubbornness, 1; efficacy created, 251; emphasizes process, 6; and frugality, 69; and happiness, v; Illegitimate child, 83; meaningfulness of intrinsic

INDEX

satisfaction, 4; mother pushes, 8; and personal hygiene, 173; as rebel, 70; relatedness for, 251; tinkering in science, 11

Einstein, Elsa: acceptance of affairs, 83; and Albert's grooming, 173; Albert in love with, 84; as older, 82; a Swabian, 70; wife of great man, 85

Einstein, Hans Albert, 83

Einstein, Kurt, 139

Einstein, Pauline: and music, 72; opposed Maric, 71; pushes, 72, 73; a Swabian, 70

Eisner, Kurt, 149

elevator experiments, 86, 87

Elise (Immanuel's sister), 201, 202

Elwood, NE, 63

emancipation, Jewish, 113

emigration, Jewish: after Crystal Night, 122; not permitted, 124; numbers, 112, 114, 117, 123; reasons against, 126; U.S. quota full, 123; women take lead, 127

Empire (*Reich*), German, 49, 80

Empire Saloon, 54

Endless Mountains, 21

enemies, Brecht's, 165

enemy alien control, 168

energy, a form of mass, 78

England: Adlers in, 122; bombing of, 230; child to, 128, 131; internment in, 132; less anti-German than America, 88; Thomas Naegele in, 133

English (farmers), 57

English (language): Eva Bromke knows, 232; Eva to practice, 237; inability to speak, 56; and older siblings, 55; Geiger's partner, 243

English (people), 62, 250

entanglements, quantum, 95

entertainments, 7

enthusiasms, Brecht's, 150

entrepreneurs, 241, 243, 250, 256

Ephrata Cloister, 19

epic storytelling, 157

epic theater, 169

Erie, Lake, 16

Estonia, 187, 212

Eustis, NE, 47–68

euthanasia, 126, 242

evils, anonymous, 227

exile, 160, 161

extravagances, American, 88

extrinsic motivtion, *See* motivation, extrinsic

factories, 64, 253

faith in God, 253, 254

farmers: bankruptcies, 155; longing to become, 211; refugees as, 211; and stress, 64; stubbornness of, 1; a way of life, 3

farming: autonomy in, 253; crops vs. animals, 253; difficulties in, 50; dreams of, 213; efficacy in, 253; freedom in, 253; intermittent rewards in, 252; intuitive skills in, 64; sharing work in, 64; in Swabia, 256

farms: collective, 190; size of, 63; in Warthegau, 192; Weiss' new, 192

Fear and Misery of the Third Reich, The, 162

fears, German, 120, 141, 233

featherbeds, 183

Federal Bureau of Investigation, 91, 92, 167

feelings: Brecht's, in poetry, 173; Brecht de-emphasizes, 170; Einstein's lack of, 84
feminists, 175
Feucht and Mutschler, 36, 39
Feuchts, 41
Feuchtwanger: Lion, 176; Marta, 175
film scripts, 160
finality, in plays, 178
Final Solution. *See* Holocaust
Finland, 161, 187
firefighters, 116
fires, prairie, 51
fitness, physical, 219
flags, Nazi red, 153
flak guns (anti-aircraft), 228, 229
flax, 183, 200
Florida, 82
focus, internal or external, 6, 7, 255
folk tunes, Swabian, 136
foods: blockade of, 154; from CARE, 217; leftover, 246; Swabian, 61; wartime, 225; wasted, 246; working for, 210
Ford, Henry, 224
forgiving, but not forgetting, 140, 141
formality/informality, 173
Fort Campbell, KY, 237
Fort Duquesne, 23
Fort Necessity, 23
Fourteen Points, 154
fractures, skull, 223
France, 152, 239, 259
Franco-Prussian War, 89
Frankfurt, University of, 165
Franklin, Benjamin, 21, 22
frankness, 70, 173
Frank, Philipp, 80

freedom: Brecht cherished, 145; Einstein criticizes American, 88
Freicorps (volunteer militia), 149, 150
French (troops): feared by Palatines, 14; invade Swabia, 13; and Geiger, 242; in Western Pennsylvania, 21, 22, 23
French and Indian War, 25, 26
Fressen, erst kommt, 156
Freud, Sigmund, 90
friendliness, 173
frugality: Biesdorf's, 247; Brecht's, 164; in Eustis, 57, 58; frees-up energy, 96; at Harmony, 34, 39; internalized, 259; in Swabia, 255; Geiger's, 243, 244; Weiss', 214; by women, 61
frustration, of motivation, 11
Führer, 113
Galati, 192
Galileo, 167
Galileo, Galilei, 82
Gandhi, 164
gardening, 246
Genetically Damaged Offspring, Law for the Prevention of, 241–42
geometries, 73, 83, 87
German (language): by cantors, 100; High, for religion, 58; High, on stage, 151; and Holocaust survivors, 140; in Iowa, 213; and Jews in America, 139; prohibited, 55, 56; by rabbis, 100; and religious Jews, 103
German-Americans, 3
German culture, 164, 165
German Day, 56
Germans: and Americans, cultural differences, 171; chain migration of, 49; demonstrate in Philadelphia, 24; Eastern Europeans take revenge on, 203; Einstein as a, 90; Jews

INDEX

committed to being, 141; ethnic, resettlement of, 191; ethnic, at war's end, 200; love authority, 150; and motivation to farm, 181; Nebraskan, 48; not ruthless, 150
Germany: Bromke defends, 239; Jews return to, 135; southwestern, 2
Germany/America, cultural differences, 171–74
Germany, East, 163; Brecht's playwriting in, 161; and Brecht's tinkering, 179; Brecht in, 178; subsidies for plays, 170
Germany, Jewish contacts with: Grete Marx, 136; Gretl Temes, 141; Heimann family, 138; invitations from hometowns, 140; Kurt Einstein, 140; Lilo Guggenheim, 135; Reinhold Naegele, 134
Germany, West, 169; homeless in, 209; male-female ratio in, 209; refugees in, 208
Gersthofen, 230
Gestapo: approval of music, 108, 109; arrest Adler, 118; get information, 120; get reports of emigration, 119; grill Adler, 120, 121; visit Adler home, 121, 122; was bad, 227
gestures, and Brecht, 167, 177
ghettos, 124
Gillman, Franz, 41, 42
Gilot, Françoise, 84
glances, German, 220
gleaning, 210
Gnosticism, 31
God, 3, 5, 215, 240
Goebbels, Joseph, 119, 158, 222, 223
Goering, Hermann, 160
goodness, ineffectual, 157

Good Soldier Schweyk, 166
Göppingen, 70, 129
Grabenstein, Christ 54
grasshoppers, 51, 57
gravestones, 171
gravity, 78, 86
Great American Desert, 48
Great Powers, 146
Grimm Brothers, 71, 223
Grimmelshausen, 162
grocery shopping, 245
Grossaspach, 13
Gruber (fourth-generation), 67
Gruber, Cora, 55, 58, 63, 64
Gruber, Glennis, 62
Gruber, Otto, 63
Grubers (ranchers), 62, 63
Guggenheim, Julius, 130, 135, 253
Guggenheim, Lilo, 129, 135
Guggenheim, Lini, 130
guilt, 185, 228
gum, chewing, 232
habits: conscientiousness, 255; conversion to intrinsic motivation, 12; learning, 245, 246, 247
Habonim, 138
half-Jews, 111
Hamburg, 184
Hangmen also Die, 164
Hanover, 206
happiness, 7, 83
harems, 174
Harmony: 32-42; charity by, 40; and Count Leon, 37, 38; Duss takes wealth, 41; egalitarianism at, 36; founding of, 32; frugality at, 39; to Pennsylvania, 35, 36; sacrifices at, 250
harvests, confiscated, 190

Hasek, Jaroslav, 166
Hauff, Wilhelm, 257
Hauptmann, Elisabeth, 151, 174
hazing, fraternity, 10
health care, socialized, 238
heart attacks, 144, 255
hearts, weak, 148, 179
Hecker, Friedrich, 101
Hegel, G. W. F., 173, 177, 257
Heil-Hitler greeting, 220
Heimann family: seeks synagogue, 138; and Swabian customs, 138
Heimann, Richard: and affidavits, 123; as apprentice, 114; at Dachau, 122; and Israel, 137
Heimann, Ruth, 138
Heine, Heinrich, 113, 132
Heisenberg, Werner, 94, 95
Helsinki, 161
Henrici, Jacob, 38, 39, 40, 41
Herrlingen, 115
Hesse, Hermann, 240, 241, 257, 258
Hessia, 140
Heuss, Pres. Theodor, 135
Hiawatha, 16
Highgate Cemetery, 171
high schools, Swiss, 74
Hildesheim, 206
Himmler, Heinrich, 124
Hindenburg, Pres. Paul von, 113, 153, 159, 160
hired hands, 63, 64, 186, 190, 192, 210, 213, 200
Hiroshima, 91, 167
Hitler, Adolf: adored, 194; assassination of, 185, 231; decides on Holocaust, 124; declares war on U.S., 134; enthusiasm for, 166, 226; German culture superior, 165; glorified, 218; greatest German, 185; head of government, 159, 160; incompetent, Jews see as, 107; lenient sentence for, 153; Munich coup, 153; not criticized, 220; popularity of, 193; rectifying problems, 106; as speaker, 184; takes power, 90; vision for Germany, 220

Hitler Youths: and Biesdorf, 222, 223; under the bombs, 229; enthusiasm for, 226; fighting for Germany, 11; group work in, 218; made compulsory, 221; Marine, 225; meetings of, 217; and political consciousness in, 226; Geiger in, 242

Hölderlin, Friedrich, 257
Holland, 205
Hollenzollern territory, 80
Hollywood, 163, 164, 166, 168
Holocaust: Brecht's poem about, 173; and Bromke, 239; German survivors not uncomfortable with Germans, 139; survivors, 117; survivors go to America, 212
Holy War, 221
home guard, 199
Homestead Act, 48
homework, 6, 9
honesty, 184, 246, 258
Hook, Sidney, 163
hopes, Jewish, 107, 116, 126, 127
Horkheimer, Max, 165
Horst Wessel song, 115
hospitals, 47, 97, 108, 145, 148-49, 198, 234, 242, 244
House Un-American Activities Committee (HUAC), 168, 169
housing: confiscated, 235; destroyed, 230; Jewish, 118

INDEX

Huchelhofen, 204
Hugenberg, Alfred, 157
humiliation, German, 154
humor: Brecht's dry, 173; Einstein's coarse, 70; Einstein's earthy, 71, 82; in Nebraska, 60; vulgar, 166
Hungary, 200
Hus, John, 20
Hutterites, 41
hydrogen bombs, 92
hygiene, personal: Biesdorf's, 217; Brecht's, 147, 173; Einstein's, 173
hypocrisy, 146
Ichenhausen, 102
illusions, positive, 249
Ilmen, Lake, 196
immigration, 117
impersonality, 169, 170
independence: of Bessarabia, 184; in farming, 59, 68, 211, 213, 214; Swabians value, 65
independent Independent, 150
Indiana, 34
Indians: drunken, 15, 16, 250; Iroquois, 16, 17, 18; Mohawk, 14, 15, 16, 249; Oneida, 16, 18; Onondaga, 16, 18, 21, 22; Plains, 47, 48; Seneca, 16; Shawnee, 20, 23; Six Nations, 18; Wyandot, 23
individualism: Biesdorf's, 219; Brecht's, 150; in capitalistic America, 224; Einstein's, 70; for religious decisions, 240
industries, damage to, 227
infants, 8
Inflation: Great, 152, 157; Jews as cause of, 107
innovations, Brecht's, 170
integration of habits, 8

intellect, 147, 218
intellectuals, American, 171, 165
Intelligence Agency, Nazi. See SI.
internalization: of frugality, 259; of habits, 8; for intrinsic motivation, 251
internal security control, FBI's, 168
interrogations, Gestapo, 212
intimacy, 82, 83, 93
intrinsic motivation, See motivation, intrinsic
intuitions, 95
invariance in physics, 76
invasions, German, 161, 192; Soviet, 187
inventors, 256
Iowa, 212
Ireland, 14, 66
Irish (farmers), 57
Iroquois Council Fire. See Council Fire, Onondaga
Isle of Man, 133
Ismail, 186, 188
Israel (first name), 114
Israel (nation), 88
Israelitische Annalen, 102
Italy, 74, 193
Jacobs (Englishman), 129
Jebenhausen, 70
Jefferson, Pres. Thomas, 32
Jewish firms, 114
Jewish Relief Organization, 118, 136
Jews: 99-141, all hated, 106; American, do not accept visits to Germany, 141; Americans reject, 141; Americas do not accept, 88; assimilated by 1933, 105; bad attitude of, 239; and Bromke, 239; committed to being Germans, 1;

Eastern European, 125, 138, 193; emancipation of, 103; fired from firms, 114; forced workers, 233; freedom and efficacy of, 252; German, 99–142; guilt for killing of, 228; Holocaust survivors, 211; hopes for change, 109; intrinsic commitment to citizenship, 252; killing of, German awareness, 125; as leftist subversives, 149; marriages with Gentiles, 104, 105, 110, 140; in Munich, 152; Orthodox, 72; Polish, discriminated against, 87; Polish, superior religious training of, 103; Posen, spoke Yiddish, 103; reduced autonomy and efficacy, 253; as religious/national group, 137; return to Germany, 253; Russian, 103; as scapegoats, 106; secular, 105; Swabian, assimilation of, 139; Swabian, integration with Gentiles, 104; Swabian, secular, 104, 105; Swiss, 90; wealthy leave villages, 101; and Warthegau, 192; and Weiser, 20. See also Einstein, Albert, and Weigel, Helene

Jim (Eustis resident), 65, 66

Jobs, Steve, 84

Johnson, Paul, 172, 179

Judaism: communal nature of, 100; as ethics and morality, 100; liberal, 138; Reform, 100

judges, 153

Jugendblut (child's blood), 122

Jungle of the Cities, In the, 155

Jungvolk (Young People), 218, 221, 225

Kaiser, 89

Kaiser Wilhelm Institute, 81

Kansas, 54

Karamischewo, 195

Karlshausen, 200, 201

Keitel, Gen. Wilhelm, 219

Kennedy, Pres. Jack, 5

Kentucky, 238

Kepler, Johannes, 76, 82

Kerensky, Alexander, 150

Kerner, Justinus, 257

Kidd, Captain, 14

killing, 88

kindness, 156

kitchens, sharing, 211

Klein Lohe, 192

Kleist Prize, 151

knitting, 183

Koch, Herman, 57, 61

Kolberg, Pomerania, 198

kosher butchering, 113

Kulm: buses come, 192; under Communists, 190; description of, 182; getting home, 190; and repatriation, 191; survivors from war, 207

Kussy, Gustav, 102

Kussy, Sarah, 102

Laemmle, Carl, 117

Lancaster conference, 21, 22

Landjäger, 115, 116

language, on stage, 151

Latvia, 184, 212

Laughton, Charles, 167, 171

Laupheim, 116, 117

laws, respect for, 113

League of Nations, 90

Lebanon Valley, 18

Lech, River, 235

leftists: Brecht's women as, 175; Einstein against tyranny of, 89; flee in 1933, 107; in Munich, 152

INDEX

legality, German preference for, 112; of killings, 125
Leningrad, 196
Lenin, Vladimir, 150, 168
Lenya, Lotte, 175
Leon, Count, 37, 38, 42
Lewin, Kurt, 172
Liebknecht, Karl, 149, 150
Life of Galileo, 179
light, speed of, 77, 86, 95
Lisbon, 132
Lithuania, 212
Livingston Manor, 14, 15, 17, 250
Livingston, Robert, 14, 17
Locarno Treaty of, 220
Logan, James, 25, 26
London, 14, 135
Long Beach Harbor, 168
Los Angeles, 163, 164
losses, Slavic, 199
Louisiana, 38
love, hypocritical, 240
luggage, 208
Lüneburg, 206
Lutheran Church, 19
Lutherans, 29, 30, 31, 250
Luther, Martin, 145
Luxemburg, Rosa, 149
lynchings, 56
McCarthy era, 92
Macheath, 174
machines, farm, 52
Madagascar, 123
Mädchen, Bund Deutscher, 105, 106
magnetism, personal, 147, 148
Mahogany, 153
Malumin, Dimitri, 188, 189
Manhattan, 138
Man Is Man, 155
Mann, Thomas, 165
manure, 182, 183
Marcuse, Herbert, 165
Maric, Mileva, 71, 82, 83, 84
marriages: Brecht's, 151; Einstein's second, 85; Einstein's view of, 83; and happiness, 7; Harmonists' 36; Immanuel and Johanna, 186; and intrinsic commitment, 3; privileged and non-privileged, 111, 112; Weiser's
Marx, Grete: contacts Walz, 120; friends of, 109, 110; goes to Gestapo, 118; leaves for America, 131; on Nazis in Stuttgart, 119; tells Adler's story, 107; visited by Gestapo, 121; visits Germany, 135, 136
Marxism: and Brecht, 169, 178
Marx, Julius, 114, 115, 116, 128
Marx, Karl, 155, 156
Marx, Leopold, 128
Marx, Liddy, 115
Marx, relatives, 132
Marx, Walter, 108
masculinity, Brecht cultivates, 145
mass, as lumped-up energy, 78
mass media, 247–48
materialism, 155
Maxwell, James Clark, 76
May Day, 115, 150, 156
meals, 52, 82
meaningfulness, 7, 9, 65
Measures Taken, The, 156
medics, 222, 223
Mein Kampf (My Struggle), 153
meningitis, 223
Mennonites, 42
Mercury, 87
merit, promotion by, 219

Messerschmitt factory, 229
Michigan, University of, 170
middle class, Brecht attacks, 146
military, dangerous to democracy, 90
military police, 235
mind, won't stop, 244
Minnesota, 167
miracle year, Einstein's, 75
missionaries, 247
mistresses. *See* affairs (sexual)
modesty: Einstein not assertive, 81; Einstein, on money, 81; Einstein, on social status, 79
moguls, movie, 164
Mohawk River, 15, 17
Moldova, 181
molestations of women, 234
money: and conservative farmers, 57; controls politics, 93; Einstein's lack of, 71; Einstein's modesty, 81; as evil, 156; extrinsic motivation for, 254; as a goal, 242; for homework, 6; Jews pay for Crystal Night, 118; used to write plays, 164
Monthly Review, 93
Montreal, 26
morale, 227
morality: and conscience, 240; not God's concern, 73
Moravian Church, 20
Mörike, Eduard, 257
Moscow, 162; Brecht avoids, 162; Brecht does not tarry in, 251; Brecht visits, 163; German goal, 193, 194; May Day in, 150; Steffin dies in, 163
Moses, 100
Mother Courage and Her Children, 162, 172, 179
mothers, relatedness by, 251

motivation, to farm, 181
motivation, extrinsic: from authorities, 251; converted to intrinsic, 11, 12, 255; Einstein's lack of, 71; and external goals, 4; grades at school, 254; Jewish concern for, 112; loss not significant, 25; and Nazi abuse, 114; and negative outcomes, 10; from parents, 252; and religious observance, 257; and selling of farms, 57; Geiger's, 244; at Zoar, 44, 45
motivation, intrinsic: from action, 9; Brecht's calling, 173; Brecht's interest in process, 177; and Brecht's trial and error, 148; Brecht cultivated autonomy for, 145; dashed by Crystal Night, 116; efficacy for, 251; Einstein's, 71; escalation of, 10; faith in God, 254; for farming, 47, 57, 65, 67, 214; freedom from extrinsic rewards by teachers, 251; hard to extinguish, 10, 248; ingredients for, 249; and instinct to grow, 8; and intermittent rewards, 252; and irrationality, 9; Jews no longer posses, 123; by Jews, to remain German, 126; negatives in developing, 9; and process, 6; and relatedness, 251; from relatedness, 240, 254; to remain Germans, 112; resistant to extinction, 252; sacrifices by communards, 250; and satisfaction, 4; secondary for Jews, 115; for spirituality, 258; for workmanship, 244; by writers, inventors, 256, 258; at Zoar, 45
Müller, Heinrich, 124

INDEX

Munich, 144; Blochs go to, 102; Brecht at university, 148; Brecht unable to settle there, 179; Einstein's high school in, 74; Einstein moves to, 72; Hitler appears, 152
Münsterer, Hanns Otto, 147, 171
Murrhardt, 135
music, choral, 136
Music, New York College of, 131
Mutschler, Hildegard, 36
Naegele, Kaspar, 133
Naegele, Reinhold: 1954 Stuttgart show, 135; background, 110; leaves for England, 132; and mixed marriage, 111; painting at Heimann home, 138; painting resumed, 134; returns to Germany, 134
Naegele, Thomas, 110, 133, 135
Nagasaki, 91, 167
Napoleon, 31, 101, 182
nationalism, German, 90
Nationalist Party, 157
Nation, Carrie, 54
NATO, 239
nature, as religion, 73
Nazareth PA, 20
Nazi Party: criticized, 220; in Munich, 152
Nazis: American, 141; at Brecht's plays, 156; burn books, 92; and degenerate physics, 92; enthusiasm for, 166; and modern music, 110; and mystical will of people, 178; people disgusted by, 194; and people who mattered, 158; popularity lacking, 185; prevent Brecht's plays, 160; quasi-religious and apocalyptic, 106; in Reichstag, 157, 159; stop Jewish doctors, 111; voting for, 158, 160; warn Naegele, 111
Nebraska: 47-68; farmers in, 252; interviewees, 47; Territory, 47; University of, 64
Neckar River, 130
negative/positive outcomes, 252
negatives, acceptance of, 249
negativity, 146, 155, 177
Neuffen, 115
Newark NJ, 102
New Jersey, 238, 239
New School for Social Research, 161, 162
New Testament, 143, 147
Newton, Isaac, 76, 82, 86, 87, 95
New York City: Brecht in, 166; Brecht rejects, 163; Karl Adler in, 136; Palatines to, 14; Reinhold Naegele in, 134; Thomas Naegele in, 133, 135
New York Colony, 14, 15, 17, 22, 25
Nobel Prize, 78
Noerdlinger, Alice, 110, 111, 132, 134, 135
non-aggression pact, 187, 193
Normandy landings, 198
norms, cultural, 12
Norway, 161
Nuremberg Racial Laws, 111, 113
obsessional motivations, 84, 126
occupation, Russian or Soviet, 181, 187-191, 202-203, 207. *See also* Germany, East
Oder River, 205
Oetinger, Friedrich Christoph, 73
officer's clubs, 237
officials fired, Jewish, 107
Ohio River, 23, 36
oil, Germany's, 222

Old Economy Village, 42
Old Testament, 146
Oneida commune, 33
onion model of personality, 172
open-ended plays, 172
opportunities, preparation for, 247
opportunity costs, 245
optimizing/satisficing, 254
Ora et Labora, 41
order: and Einstein, 73, 82; and German Jews, 137, 138; Germans prefer, 153
Organisation Todt, 226
organs (musical), 101
Orthodox Jews, 100
Osterberg, 102
outside contacts, 250
Owen, Robert, 36
pacifism: of *All Quiet on the Western Front*, 158; Brecht's, 143, 147, 176; Brecht's, and houseflies, 156; Brecht's, in East Berlin, 178; Brecht's scripts, 164; Christ's, 162; Einstein's, 88; Einstein's, limited, 90, 91; of Independent Social Democrats, 149; of *Mother Courage*, 172; in Swabia, 88
paintings, abstract expressionist, 134; German expressionist, 134
Palatines, 14, 15, 17
Palestine, 87, 128. *See also* Zionism
Papen, Franz von, 159, 160
parades and rallies, 159
paradigms, scientific, 76, 252
paraplegics, 7
parents, 8, 9, 12, 72, 259, 260
Paris, 132, 195
partisans, 195
pastors, 109, 212, 213
patent office, 75, 76

patents, 257
patriotism, 147, 164
Peace Corps, 5
peasants, free, 190
peasants, Polish, 192
Peasant War of 1525, 32
Pemberton, Israel, 25, 26
pennies, picking up, 242
Penn proprietors, 21, 25
Pennsylvania Assembly, 21, 24, 25
Penn, Thomas, 18
Penn, William, 17
perfectionism, 82, 255
permissiveness, 259
personalities, 155, 169, 170
personality, Type A, 254
Philadelphia, 23, 43, 237
photons, 78
Picasso, Pablo, 84
Pietism: Brecht influenced by, 145; and Einstein, 70, 74, 96; and emigration to Bessarabia, 182; and feelings, 173; and intrinsic motivation, 257; and Rapp, 29, 34; and usefulness, 171; Weiser's, 18, 19
Pittsburgh, 32, 36, 40, 245
Planck, Max, 78, 252
Platte River, 48, 53, 56
Ploesti, Romania, 222
plundering, by Americans, 234
poetry, 144, 176
pogroms, 101, 112. *See also* Crystal Night
Poland: and Gestapo, 136; invasion of, 132, 186, 187
politicians, ineffective, 153
Pomerania, 200
pomposity, 165
Pontiac's Conspiracy, 26

INDEX

Portland OR, 42
Portugal, 131
Posen, 103, 192
Potsdam, 209
poverty, 176
practicality, 171
Prague, 81
Prahova, 192
prairies, 50
praise, 171
praxis, 155, 165, 178
prejudices, 88
pressures. *See* motivation, extrinsic
pretentiousness, 167
price tags, in America, 164
pride, German, 154
Princeton, NJ, 69, 82, 91
principles, adherence to, 137
prisoners of war, 133, 204–5
prison, Stanford, 10
privacy, 173
Private Life of the Master Race, The, 162
probabilities, 94. *See also* quantum mechanics
process: and athletes, 6; gives meaning to life, 241; and intrinsic motivation, 255; and not goals, 176
professors, 109
Prohibition, 59
prohibitionists, 54
propaganda, 56, 147
Propaganda Ministry, 119, 184
propriety, 147
prostitutes, 151, 236, 174
Protestant Reformation, 255
Protestants, 14
Prussia, 160
Prussia, East, 132

Prussian Academy of Science, 81
Prussians, 49, 80, 182
psychopaths, 176
public works, 184
punishments, 4; corporal, 183, 258
Putsch (coup), 153
Quakers, 21, 23, 24, 25
quantity buying, 245
quantum mechanics, 73, 84, 252
quantum theory, 78, 94
quarter Jews, 111
quotas: university, for Jews, 99; U.S., 117, 123, 161, 212
rabbis, 99
race, Jewish, 106
radioactivity, 78
Radio, People's, 184
radios, 128, 184, 228, 229
railroads, 49, 50, 51, 65, 150, 196, 207, 225, 230
rain, 51
ranching, 63, 64
Rapp, Frederick, 37, 38
Rapp, Johann Georg: to America, 32; attracted to Mutschler, 36; birth, 29; death, 39; rejects Lutheranism, 30; and sacrifices, 33; and Second Coming, 250; strictness of, 44; to Washington, 33
reactance, psychological, 2
Reading PA, 17
realism: social, 163; unromantic, 145
rebelliousness, Brecht's, 144, 147, 148; Einstein's, 70, 74, 75
reconciliation, 154
records, genealogical, 191
Red Army, 195, 196
Red Cross, 206
Reformed Church, 19

refugees: from annexed lands, 201; from bombing, 228, 230; recover slower, 211; taking in, 201
Reich, Bernhard, 175
Reich, Fourth, 138
Reich, Third. *See* Nazis
Reichstag: burns down, 107, 160; Einstein at, 90
relatedness: of farmers, 47; and instinct to grow, 8; and internalization, 9; for intrinsic motivation, 249. *See also* parents
relationships, interpersonal. *See* relatedness
relatives, German, 162
relativity: general, 78; special, 78, 86
relaxation, 172
religion: and conscientiousness, 258; Einstein's, 73, 74; in Eustis, 58; Jewish, 83; at school, 254
Remarque, Erich Maria, 158
repatriations, 191
repertory system, 170
reputations, 258
research institutes, 257
resettlement of Jews, 123, 124
residence permits, 207, 208
resistance, German, 166, 194, 233, 235
Resistible Rise of Arturo Ui, 165
respect, Germans seek, 220
retreat, German, 231
Revelation, Book of, 31, 32
revolution, advocacy of, 168, 169; American, 24;, future, 178; Nazi social, 219
rewards, external. *See* motivation, extrinsic
rewards, extrinsic. *See* motivation, extrinsic

rewards, intermittent, 252
rewards, internal. *See* motivation, intrinsic
rewards, intrinsic. *See* motivation, intrinsic
Rhineland, 220
Rhine River, 14, 242
Riesman, David, 2
Riga, Latvia, 139, 198
rigidity, German, 137
Rise and Fall of the City of Mahagonny, 156
robberies, by soldiers, 235
rocket launchers, 204, 232
Romania, 181, 184, 200, 215
Romanian (language), 184
Romanians, 181, 184
Rommel, Gen. Erwin, 219
Roosevelt, Pres. Franklin Delano: and airplane production, 222; and Civilian Conservation Corps, 219; Einstein's letter to, 91; Einstein visits, 69; Naegele on, 133; writes Stimson, 203
Rothschild, Ray, 140
Rottenburg, 241
Royal Israelite Superior Church Authority, 99
Rudel, Col. Hans-Ulrich, 219
Ruhr, French invade, 90, 152
rumors, 127
Russia: America against, 227; Czarist, 149; Czar of, 181; German invasion of, 221; invades Bessarabia, 189; total war by, 223. *See also* Soviet Union.
SA (*Staatsabteilung*). *See* Storm Troopers

sacrifices: Brecht's, 160; at communes, 33; by Conrad Weiser, 15, 16; of farmers, 47, 53; at Harmony, 32, 35, 38, 39; at long-lived communes, 45; Nazis emphasize, 218; strengthened Einstein's motivation, 96, 97; strengthen intrinsic motivation, 10
Saint Joan of the Slaughterhouses, 157
sales, buying at, 245
saloons, 59
Salvation Army, 157
Santa Monica, 164
Sarah (first name), 114
Sarajevo, 89
satisfactions, in farming, 65. See also motivation, intrinsicintelligence
satisficing/optimizing, 254
Sausage Day, 61
Saxe-Weimar, Duke of, 36
scalps, 23, 24, 25
Scandinavia, 161, 163
Scandinavians, 62
Schelling, Friedrich, 257
Schenectady, 15
Schiller, Friedrich, 171, 257
Schmeekle, John, 49
Schmidt, Hans, 225
Schoharie Creek, 15, 16
scholars, as butchers, 167
Scholl, Hans and Sophie, 194
schooling, 8, 12, 55, 64
school teachers, 109
Schweinfurt, 192
Schweyk in the Second World War, 166
science-religion conflict, 73
scientists, 256, 257
Scotch-Irish, 17
Scotland, 133
Scouts, Cub and Boy, 218

SD. *See Sicherheitsdienst*
searchlights, 228
Second Coming, 31, 36, 37, 74, 250
secret pact, 187
sects, in Swabia, 258
self-determination. *See* autonomy
Separatists, 30, 31, 32, 43, 250
Serbians, 89, 202
Sermon on the Mount, 162
Seven Hares, 151
Seven Swabians, 71
sex (relations): contacts with German women, 235, 236; diminished intrinsic motivation, 250; illegal, 113; premarital, 219
sex (roles), and alcohol, 66
sexism: Brecht's, 173, 174
Shakers, 41
shame, 154
share-croppers, 213
sharing, 217, 218, 224
sheep, black, 217
shelters, air raid, 229
Shenandoah Valley, 21
Shickellamy, 18, 21
shocks, electric, 10, 11
shoes, 162
shortages, postwar, 209
shtetls, 103
Siberia, 190
Sicherheitsdienst (Nazi Intelligence Agency), 119, 165, 193, 194
Siddhartha, 241
siding, aluminum, 243
Sie/Du, 173
Silesia, 200
Silicon Valley, 257
Simone, 164
simple-living. *See* frugality

sincerity, 172
Sinclair, Upton, 155
singing, 151
single-mindedness, 3, 93, 95
slave laborers, 234
Snow, C. P., 93
Social Democrats: attacked by Communists, 148; attacked by right, 149; in Bavaria, 149; to concentration camps, 233; defend Weimar Republic, 159; massacre workers, 156
socialism, 92, 93
Socialists, 107, 149
Social Research, Institute for, 165, 166
society, problems in, 170
sod houses, 50, 55
Solzhenitsyn, Aleksandr, 199, 240
South Africa, 130
Soviet Man, New, 155
soviets (councils), 149
Soviet Union: Americans sympathized with, 215; Cold War with, 168, 212; Finland defends against, 161; German invasion of, 123; invasions by 187; to pay Bessarabians, 191
Soviet Zone of Occupation, 207
Spartacus, 149
speaker, Hitler as, 220
spending, difficulties, 248
spinning wool, 183
spirituality, 257
sponsorships: Heimann family's, 123; U. S. law for, 117; Weiss family's, 212
spooky actions at a distance, 96
SS (*Schutzstaffel*): in charge of Gestapo and camps, 119, 120; and fight to death, 232; in Kulm, 191; *and Volkssturm*, 226

stab in the back, 107, 110, 231, 232
stage language, 156
Stalin, 187, 221
Stalingrad, 194
Stalinism, 163
Stanford, Sen. Leland, 51
Stanislavsky System, 170
starlight bent, 87
Stauffenberg, Claus von, 166
stealing by Brecht, 175
Steffin, Margarete, 161, 162, 163, 174
steppe, 182
stereotypes: of German drinking, 67; of German stubbornness, 1, 239; of Indians, 16; of Swabian stubbornness, 2; of threatening German, 87
sterilizations, 242
Stern, Otto, 94
stimulation, 4
Storm Troopers: appeal to youths, 158, 159; boycott by, 112; leaders murdered, 119; stand up to Communists, 159; violence against Adler, 108; during wartime, 226
Strategic Services, U.S. Office of, 165
Streep, Meryl, 179
strength, Hitler communicates, 185
strictness, for Weiss, 183
stubbornness: by children of Swabian immigrants, 142; E*igensinn*, 2; Germans stereotyped, 1; leading to happiness, 2; negative, 2; positive, 2
students, Nazi, 112, 113
Stuttgart, 13; and Bach's music, 136; bombing of, 223; Crystal Night in, 117; deportation of Jews, 139; Guggenheim house in, 131; Nazis film deportation, 124; "our

INDEX

Jerusalem," 100; public opinion in, 193, 194; secular Jews in, 139
subsidies, theater, 170
Succoth, 121, 122
suicides, 161
suitcases, 115, 228
Sundays, 58
Super Shoppers, 245
support, public, 194
surrenders, 204, 233
survival, 195
Swabia: alcohol in, 47; apocalyptic views in, 92; asceticism in, 69; Bavarian, 101, 102, 228; children work in, 53; conscientiousness in, 73, 138, 143, 243, 244; criticism in, 143, 165, 171; cultural emphases, 256; egalitarianism in, 79, 80, 165; farming in, 63; frankness and contentiousness in, 70, 88, 173; frugality in, 11, 163; having it, not showing it, in, 39; humor, 60; importance of behaviors, 169; importance of God, 73; independence in, 71, 79, 241; individualism in, 218; inheritances in, 79; intrinsic motivation in, 256; marital strife in, 83; modesty in, 79; moralism in, 143; music in, 35, 72; negativity in, 146, 177; pacifism in, 88; peasants in, 252; people get along, 16; perfectionism and criticism in, 82; religious courts in, 39; religious values in, 143; resistance to authorities, 258; self-reliance in, 34; simple living in, 135; spirituality in, 18, 257; stubbornness in, 2; values in, 65, 131; Weiss family returns to, 209; women distract in, 85; work ethic in, 241
Swabian (language), 55, 140
Swabians: actions by, 71; bumpkin, 151; and Palatines, 14; emigration to Bessarabia, 181; Valiant, 252
Swabians, Seven, 223
swearing, 58, 60
Sweden, 161, 162
swing youths, 219
Switzerland, 80, 113, 127, 169, 184
synagogues, 100, 137
tanks, 178, 197, 251
tattoos, 192, 205
Tax, Reich Flight, 132
teachers: in Bessarabia, 184, and burnout, 251; caring, 8; controling, 9; and corporal punishment, 183; einstein's 70, 74; as government employees, 109, 120; Rapp on, 30; seduced by exrinsic rewards, 260; Sidhartha's 241; and strictness initially, 259, warmth of 259;
Teachers, American Federation of, 93
teaching-learning plays, 156, 251
Technology, Swiss Institute of, 71, 74, 82, 252
Teedyuscung, 25
Temes, Gretl, 106, 127, 139, 141
terror, Red/White, 150
theater, American, 169, 170
Theresienstadt, 128, 203
thinking, and Brecht, 170, 171
Thirty Year's War, 162
Thomas, Rep. J. Parnell, 168
thought experiments, 77, 86, 95. *See also* trial and error
threats, Nazi, 91, 121

Threepenny Opera, The, 145, 151, 156, 174, 175
time/space, not fixed, but elastic, 76, 78
Times Square (NYC), 133
Times, The New York, 88
tinkering, 75, 76. *See also* trial and error
Tocqueville, Alexis de, 88
top dog, oneself as, 149
toughness, Brecht's, 147
tourists, soldiers as, 236
traits, Einstein's Swabian, 70
Transylvania, 184
trial and error, 11, 148, 300. *See also* thought experiments.
Trial of Lucullus, 162
Tripartite Pact, 193
trust, 57, 70, 121
Tübingen, 223
tugs-of-war, 167
Tulpehocken, 17, 21, 24
Tunisia, 194
Turks, 71, 181, 184
Uebele, Kate, 52
Uhland, Ludwig, 71, 257
Ukraine, 181, 189, 199
Ulbricht, Walter, 162, 178
Ulm, 42, 72, 115
Un-American Activities, House Committee for, 92, 161
unbudgeable, Einstein as, 93
unemployment, 158, 219, 220
unfaithfulness (marital), 85. *See also* affairs
unified field theory, 95
Union Pacific Railroad, 49
University of Pittsburgh, 245
usefulness, 171
Valiant Swabian, 71, 73
Valkyrie, 166

Vanderbilt, Cornelius, 40
vanity, 82
venereal diseases, 234
Versailles Treaty: and demilitarization of Rhineland, 220; expectations dashed by, 154; and French-German enmity, 152; Jews cause, 107; keeps Germany down, 218; Nazis to avenge, 159
Vienna, 179, 231
Vietnam War, 239
Villon, 175
violence, street, 90
violence, youth, 159
Virginians, 21, 23
Vladivostok, 162
Vogel, Henrietta, 49
Volk (people or society), 106, 218, 222
Volkssturm (home guard), 226
vulgarity, 60, 71, 82, 140, 166, 170
Wabash River, 34, 35
Waffen-SS, 196
wagon trains, 201
Waldoboro, Maine, 15
Waldo, General Samuel, 15
Waldorf-Astoria Hotel, 69
Geiger, Rudy, 11, 241, 254, 255
Walz, Hans, 120, 121
war, enjoy the, 231
War, Franco-Prussian, 80
Warsaw (city), 199
Warsaw (duchy), 182
wars, futility of, 162
war, solutions to, 89, 90, 92
Warthegau, 192, 199, 204, 214
War, Total, 223
War, Vietnam, 224
Washington, Col. George, 23
Washington, D.C., 168

INDEX

water, 50, 52
Waterloo, 205
wave-particle duality, 78
weapons, secret, 226
Weigel, Helene: and Brecht's affairs, 174; Brecht's financial responsibility for, 163; Brecht in bed with, 151; a Communist, 155; directorship of Berlin theater, 169; notices agent, 168; put on social events, 175
Weimar Republic, 107, 148, 178
Weiser, Ann Eve, 17
Weiser, Conrad, 13–27; death of, 27; at Ephrata Cloister, 19; interpreter and ambassador, 18; at Lancaster conference, 21, 22; to Livingston Manor, 14; loses ambassadorship, 25; marries, 17; to Mohawks, 15; and Ohio Indians, 23, 24, 25, 26; to Pennsylvania, 17; as Pietist, 18; process of commitment, 53; rejected by New York Colony, 22; and relatedness, 249; stubbornness of, 1; with Zinzendorf, 20
Weiser, John, 13, 14, 15, 17
Weisersdorf, 15, 250
Weiss, Immanuel: drafted, 194; extrinsic motivation and honesty, 258; intrinsic motivation of, 252; single-mindedness of, 3
Weiss, Immanuel and Johanna, ancestors of, 181
Weiss, Johanna, 183, 199–203
Weizmann, Chaim, 87, 88
welfare payments, 219, 220, 241
West Front, 203–4
West Point, 219
whiskey, 34, 54, 59, 60
White Rose, 166, 194

willpower, 242
Wilson, Pres. Woodrow, 154
wine, 58, 183
Winter Help, 241
women: and Brecht's writing, 174, 175, 176; Einstein's view of, 85; influence American soldiers, 236; in Nebraska, 52, 53; Polish, protect Germans, 203; raped, 202, 203
Women's Christian Temperance Union, 59, 60
word, keeping one's, 122
work: in bombed cities, 230, 231; Brecht's inability to, 163; intrinsic motivation and happiness, 7; intrinsic motivation and productivity, 8; as refugee, 210; and relaxation, 172
workaholism. *See* work ethic.
workers: Brecht emulates, 147; forced, 231; foreign, 235; Polish, 199; uprising of, 178, 179
work ethic: avoiding wasted time, 247; becoming entrepreneurs, 241; Brecht's, 164; Einstein's, 70, 97; Laughton's, 167; and loneliness, 62; in Nebraska, 57; Geiger's, 243, 244; Weiss family and, 214
work, modern, intrinsic motivation for, 257
work, tool-and-die, 223
world, end of, 10
world government, 89, 92
World, Manifesto to the Civilized, 89
World War I: Einstein during, 86; in Nebraska, 55, 56; and painting, 111; recalled, 221, 222; shame from, 154; triggers for, 89

World War II: and Biesdorf, 218-19, 221-24; and Brecht, 161, 167; and Bromke, 227-33; and Einstein, 91, 96; and Geiger, 242; and Jews, 123-25, 128, 129, 132-34, 139-40; and Schmidt, 224-27; and Weiss, 186-207.
wormholes, 96
wounds, 198
writing, 148, 257
Württemberg, 31, 32, 43, 49
Württemberg, Duchy of, 101
Württemberg, King William II of, 79
Yiddish, 87, 141
youths, rebellious, 154
Yugoslavia, 193
Zeigarnik, Bluma, 172
Zeigarnik effect, 172
Zelienople, Harmony Museum at, 42
Zero Time, 208
Zinzendorf, Count, 20
Zionism: becomes appealing, 108; cultural vs. nationalistic, 87, 88; and Einstein, 87; and German Jews, 87, 137
Zoar commune: celibacy at, 44; democratic, 43; end of, 45; Harmony helped, 41; individualism at, 44; not strict, 250; origin of name, 42; work ethic at, 43
Zoff, Marianne, 151
Züllichendorf, 201, 207
Zurich, 79, 179

THE AUTHOR

George F. Wieland holds a B.A. in psychology from Stanford and a Ph.D. in social psychology from the University of Michigan, Ann Arbor. He has taught and conducted research at Michigan, Vanderbilt, and Guy's Hospital Medical School in London, England. He has written several books on hospital and organizational cultures. His latest book on German immigrants is *Celtic Germans: The Rise and Fall of Ann Arbor's Swabians*. Originally from New York City, he currently lives in Ann Arbor.

Made in the USA
Middletown, DE
18 April 2016